Fine Homebuilding

P9-DEJ-033

TIPS & TECHNIQUES

FOR BUILDERS

Fine Homebuilding®

TIPS & TECHNIQUES

FOR BUILDERS

Cover Illustration: Robert La Pointe

© 1988 by The Taunton Press, Inc.
All rights reserved.

First printing: 1988
Second printing: 1990
Third printing: 1994
Fourth printing: 1995
International Standard Book Number: 0-942391-09-8
Library of Congress Catalog Card Number: 88-50563
Printed in the United States of America

A Fine Homebuilding Book
Fine Homebuilding® is a trademark of The Taunton Press, Inc., registered in the U.S. Patent and Trademark Office.

The Taunton Press, 63 South Main Street, Box 5506,
Newtown, CT 06470-5506

Table of Contents

Introduction

Fine Homebuilding is a magazine written for builders, by builders. Its pages present in-depth articles about all the construction processes needed by a person standing in an empty lot who wants to get from the footings to the ridge vent. But there's another kind of article in every issue: the diminutive pieces that appear in the Tips & Techniques column. They are collected here for the first time.

The Tips column, like late Friday afternoon, is where builders kick back and share their hard-earned victories over job-site conditions (on a cold day, keep your fingers nimble by using nails warmed in a bucket over a fire made of offcuts). And like any good conversation, this exchange is always lively. For instance, when one builder's method for framing a 45-degree corner hit the newsstands, it drew a half-dozen more ways of doing the same task, each a little simpler than the last.

Any seasoned builder can tell you that a key to survival is the ability to improvise. Say your helpers are mired in traffic and you've got to lay a tongue-and-groove subfloor, raise a framed wall or snap a bunch of chalklines. Where do you get the extra hands? The builders who have contributed to the *Fine Homebuilding* Tips column have been there, and you'll find their answers in this book, along with hundreds more pieces of valuable advice.

In the eight years since the inception of the column, I've received plenty of help in separating half-baked tips (which don't see print) from the fully-baked ones you see here. My gratitude is aimed in particular at John Lively, Betsy Levine, Deborah Fillion, Mark Feirer, Kevin Ireton, Chuck Lockhart and Paul Spring. But most of all, my thanks go out to the tipster builders who've taken the time to send in their ideas. Please continue to do so.

—Charles Miller, Tips & Techniques editor/illustrator

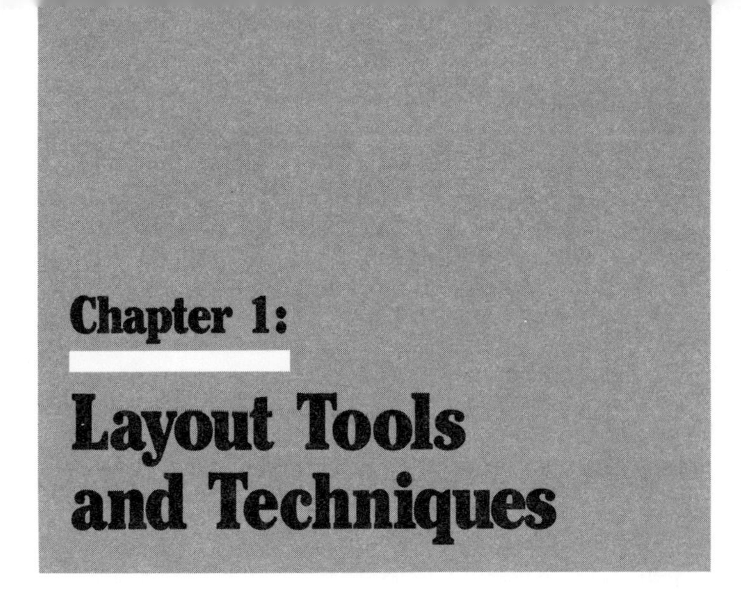

The slip stick

This versatile time-saver uses a simple story-pole principle. I call it the slip stick, and I use it to measure distances between floor and ceiling without running up and down ladders or bending my tape into hard-to-read compound bends.

The basic components of the slip stick are an old tape, a few pieces of 1x stock and whatever hardware you have around to make it work. The first component is a 7-ft. U-shaped housing, as shown in the drawing on the facing page. The second is a 6-ft. runner that fits inside the housing, where it can slide up or down. The runner has a portion of a tape-measure blade screwed to it. The portion of the tape you'll need is from the 7-ft. mark or so (attached to the top of the runner) down to about 13 ft. If you want to get fancy you can cut a concave curve into the runner, so the edges of the runner and tape will be flush.

Near the top of the assembly, I hold the two components together with a metal band that doubles as a reference point. It has to be precisely placed so that the tape at that point reads the exact length of the stick with the runner retracted. At the bottom of the runner I bolt a large washer on each face to keep the runner in the housing.

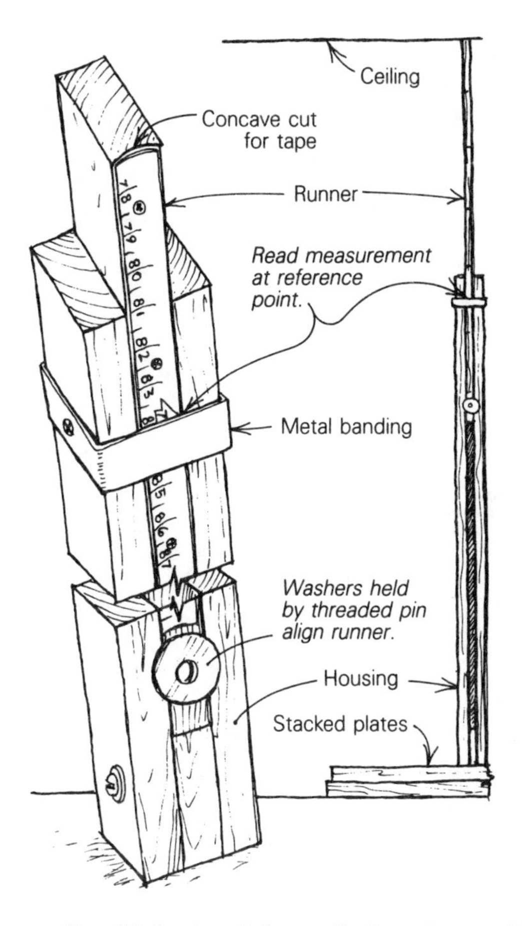

When my slip stick is closed, I can slip it under a ceiling of minimum height. With the 6-ft. runner extended, I can measure ceilings up to 13 ft. high. When I measure for a new wall in an old space, I cut the top and bottom plates and set one atop the other. With my slip stick on top of them, I can read the exact stud length without having to subtract the thickness of the plates. *—Sam Yoder, Cambridge, Mass.*

Scratch scriber

Don't throw out deformed Phillips-head screwdriver bits. Grind them to a sharp point, and use them in a pencil scriber for scratch scribing. *—Tom Law, Davidsonville, Md.*

Survey savvy

I build custom homes in the mountains, and laying out a foundation when none of the points are at the same elevation (sometimes differing up to 20 ft.) is a time-consuming, frustrating job without an expensive transit or an elaborate system of stepped batter boards.

The tip here is to have the surveyor, with his $20,000 laser transit, set the corners of the house at the same time he locates the initial property lines. The surveyor I use charges only $20 for this extra service, and it saves me what can amount to an entire day's work.

—Eric Carter, Green Valley Lake, Calif.

Dryline spool

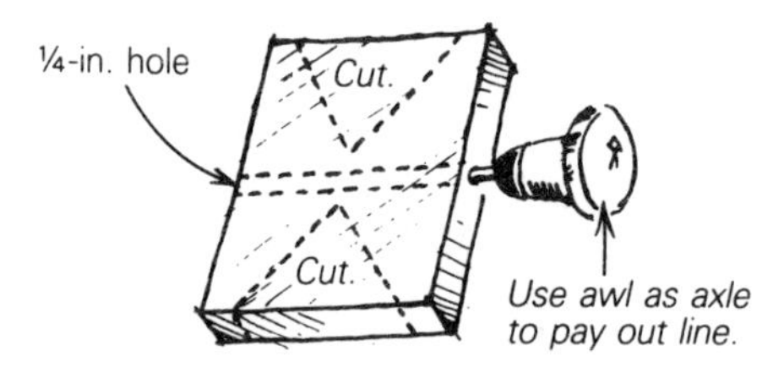

The best type of dryline spool that I have found is homemade, fashioned from a piece of ½-in. thick acrylic. I cut V-shaped notches out of the acrylic blank, as shown in the drawing, and then drilled a ¼-in. dia. hole edgewise through the center of the spool. To pay out the line, I place an awl in the hole, where it acts as an axle as I walk out the line.

Acrylic is remarkably strong. My stringline spool has survived a 50-ft. drop onto concrete, and being run over by a dumptruck. Cost of the acrylic blank: $2 per lifetime.

—Jeffrey D. Taylor, Corvallis, Ore.

Durable chalklines

Chalk layout lines are invariably erased by brooms, rain and shoes—especially on concrete. If you like to do your layout well in advance of the work it will guide, spray the lines with polyurethane varnish or some other clear finish. The lines will stay put until you need them.

—Dennis Lamonica, Panama, N. Y.

Arch trammel

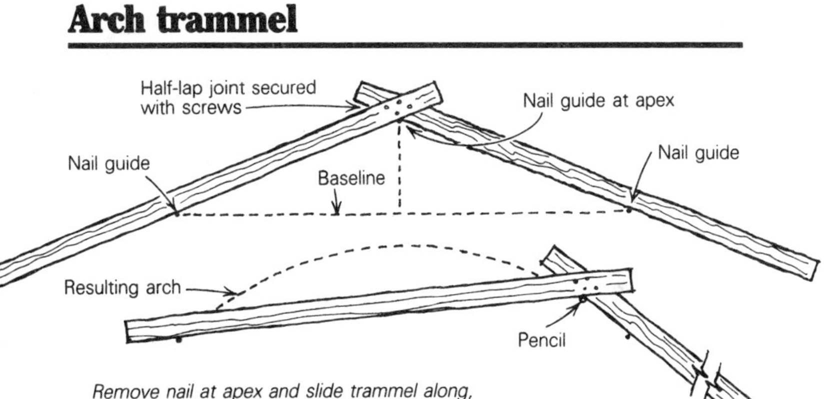

Remove nail at apex and slide trammel along, pivoting on nail guides, to describe arc.

Here is a technique that I have found handy for laying out wide, shallow arches. I discovered it at a shipyard, where boatbuilders use it for laying out the rounded camber of boat decks.

On a clear, flat, wood surface, such as a subfloor, draw a straight line as long as the arch will be wide, as shown above. Drive a finish nail at each end of this baseline, so that about 1 in. of the nail's shank projects above the floor. Find the center of the baseline, and draw a perpendicular line extending up from the center, making an inverted T. Decide the height of your arch, measure along the perpendicular line from the baseline, and drive a third finish nail at the apex.

Now find two straight 1x boards, each one a little longer than the length of the baseline. Snug one board against one base nail and the apex nail, and lay the other board against the other base nail and the apex nail. Where the boards overlap, mark them for a half-lap joint. Then cut the joint and screw the boards together.

Pull out the apex nail, and hold a pencil in its place. Now slide the boards along the baseline nail guides from right to left to mark the arch. As a variation, you can mount a router at the apex and cut or mold arches of any size.

—Jerry Azevedo, Corvallis, Ore.

Stabilizing plumb bobs

As a construction millwright, I frequently need to establish very accurate plumb lines for machinery alignment in places where a transit is impractical. I rely on my 24-oz. brass plumb bob. When it's not necessary for the end of the bob to be visible, I immerse it in a bucket of oil to reduce motion from vibration and wind.

To avoid the problem of winding and unwinding line, I replaced the string in an empty chalkbox with 50-lb. test braided nylon fishing line and tied the bob to the end; it makes a convenient reel. *—Dave Walter, Rossville, Ill.*

Butterfly spool

To avoid losing a ball of string when you're walking a plate, laying out a hillside or stringing a ridge, wrap your line around a butterfly spool. You can cut one out of plywood, or nail a couple of short pieces of lath together. To play out the line, hold the string instead of the spool. The butterfly will unwrap in jerks and jiggles, never getting too far away from you. If it does happen to wander, just give it a jerk and it will fly right back to you. *—Robert Rix, Arlington, Va.*

Foundation line markers

When I string lines for a foundation, I use alligator clips to mark the outside dimensions of the building. These clips are available at any electrical-supply house and even at some large hardware stores. They're easy to see, and I can quickly move them as I adjust the lines into square. I usually keep about a dozen clipped to my apron as I work.

—Charles Fockaert, Eureka, Calif.

Limeless lines

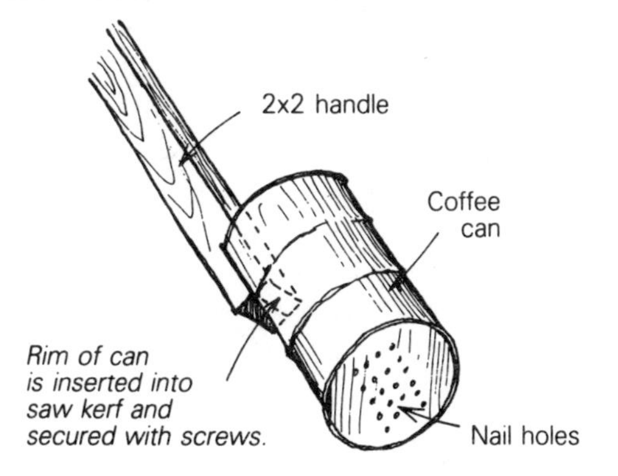

Rather than scatter handfuls of lime to mark out excavations, I use flour. It's cheaper than lime, less toxic and very available. I put the flour in a coffee can that has a grid of nail holes punched in the bottom. To make the can easy to control, I insert the top rim of the can into a kerf cut in a 4-ft. to 5-ft. 2x2 handle. Screws secure the can to the handle.

To use the tool, I let it hang loosely in one hand with the can about 6 in. from the ground. I run the handle along my dry lines, tapping it with a block or hammer to sift a line on the ground worthy of a football field.

—Eric Carter, GreenValley Lake, Calif.

Pencil scribing

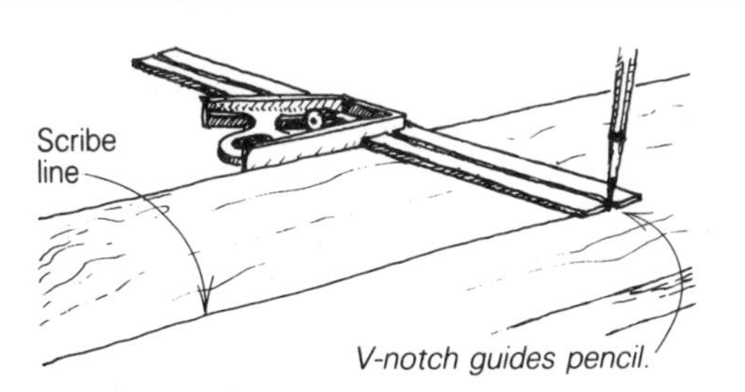

I file a small, V-shaped notch in the end of my combination-square blade to guide the tip of my pencil when scribing straight lines for ripping to width. The point stays put as the square is moved along the edge of the work, and makes a quick, accurate scribe at the width I need.

—Paul Dostie, Brunswick, Maine

Laying out a cone

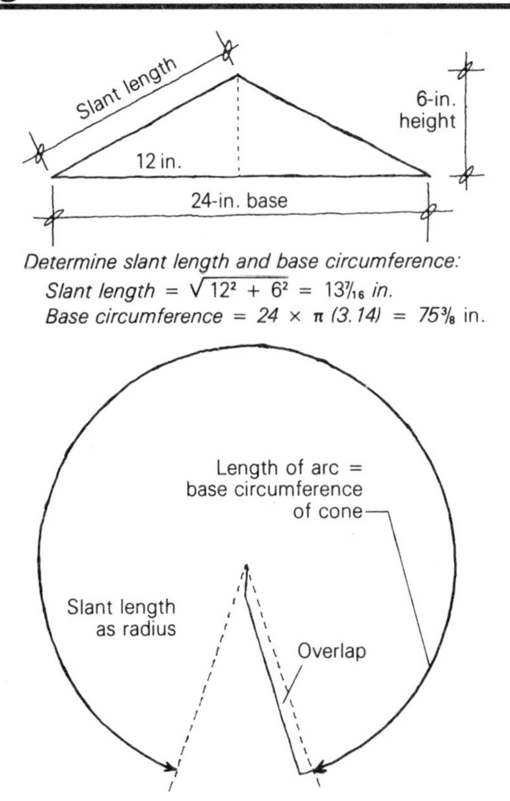

I have been intriqued with this shortcut which you may find of interest. Cones have several uses, such as sheet-metal chimney caps, but the layout of a cone is puzzling to some craftsmen. Here is a fast method that doesn't require higher mathematics. The example above shows how to lay out a cone 6 in. high with a 24-in. dia. base.

Using the slant method as radius, scribe a circle on the material, and snip. Then measure the base circumference of the cone on the circle, adding at least a ¾-in. overlap for screws, and cut the wedge.

—William A. Julien, S. Chatham, Mass.

It's a fine line

Many times timber framers need to lay out long, accurate cutting lines. A standard string in a chalk reel yields a fat, sloppy line, so I went looking for a replacement that would give me a finer line. After some research, I settled on 20-lb. test braided nylon fishing line. I find that it holds chalk very well, and it gives a crisp, clean line. If I need an even finer line, I snap it in the air a time or two before I lay it on the work.

I bought 140 yards of the braided line for about $7—enough for everyone in the shop. The only tools that I've seen that give a comparable line are Japanese ink lines, which cost about $50. ***—Tom Baker, Blissfield, Mich.***

Riser and tread marking gauge

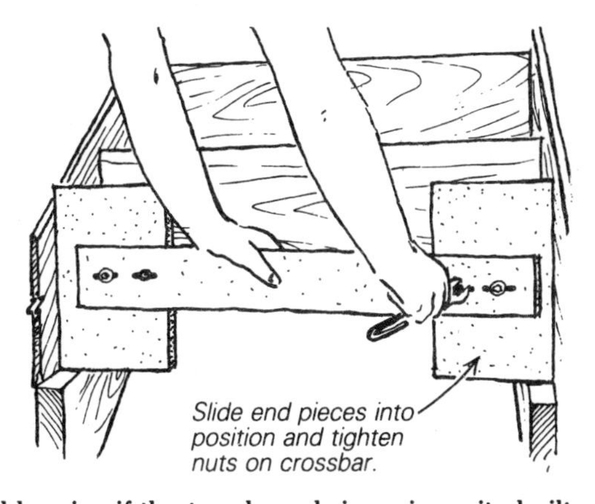

It would be nice if the treads and risers in a site-built stair could all be the same, but the vagaries of wood-frame construction usually mean that uniform lengths and 90° angles are as often exceptions as they are rules. To get accurate measurements for treads and risers that vary by just a little bit, I use the marking gauge shown in the drawing. It consists of two end pieces and a crossbar, all made of ½-in. particleboard. The end pieces are secured to the crossbar with nuts, washers and four ¼-in. flathead machine screws, which protrude through oblong slots in the crossbar. The slots allow the end pieces to be moved in and out until they make a snug fit with the skirt boards—even if they are a bit out of square. Once the fit is right, I tighten the nuts and transfer the entire gauge to the riser or tread stock, where I can mark the exact layout without making any tedious measurements. ***—James M. Westerholm, Seattle, Wash.***

Blueprint frame

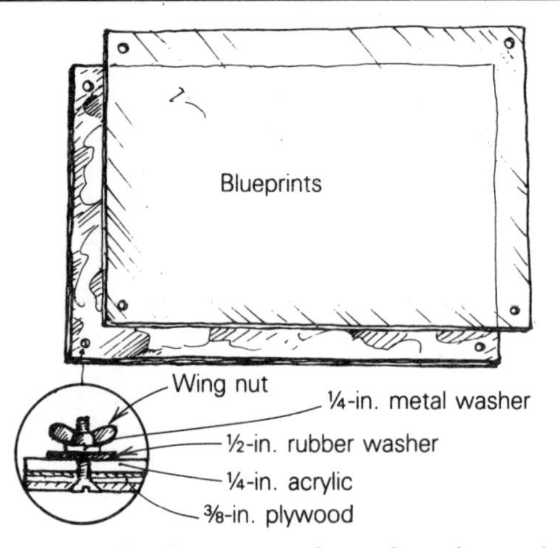

If you do construction layout, you know how frustrating it can be to decipher a set of working drawings that's been soaked by rain or blown by the wind into a mud puddle. To avoid this problem, I protect the prints with the acrylic and plywood frame shown in the drawing. I make the frame a little larger than the job drawings, and I secure the pieces at the corners with machine screws and wing nuts. I figure the effort this frame takes is worth it, for someday it will probably save me a costly error caused by some faded-out dimension. ***—David Pearson, Crane, Mo.***

Batter-board stringlines

I use the method shown in the drawing above to fasten my building lines to the batter boards. First I nail my batter boards roughly ¼ in. above my string height. Next, I mark the string height on the face of the board, and use a handsaw to cut a kerf down to my mark. The string will hold fast if it's lapped over itself in the kerf.

To make sighting easier after the string is removed, I mark a large arrow on the inside face of the board. Not only is this method easy and accurate, but it also dispenses with exposed nails, which can be bent out of alignment or even cause accidents. *—Robert J. Meehan, New City, N. Y.*

Plumbing posts

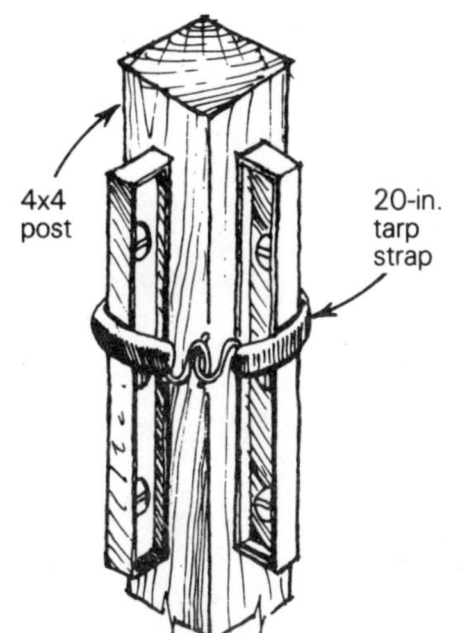

When I set posts, I find it awkward to plumb first one face, then the one next to it. One side is always getting a little out of adjustment, so I use two levels strapped to adjacent faces of the same post, as shown. For a 4x4 post, a 20-in. tarp strap is the right length to secure my two levels. Now I have enough hands to set the post, and my level isn't on the ground when I need it. *—Patrick Lawson, Sooke, B. C.*

Locating studs

If you need to locate a stud in a stick-framed wall, remember that most electricians are right-handed. Find an outlet, and tap the wall directly to its left. The odds are in your favor that the stud will be there, and you can measure away from it in 16-in. increments to find other studs.

—Art McAfee, Edmonton, Alta.

A better plumb bob

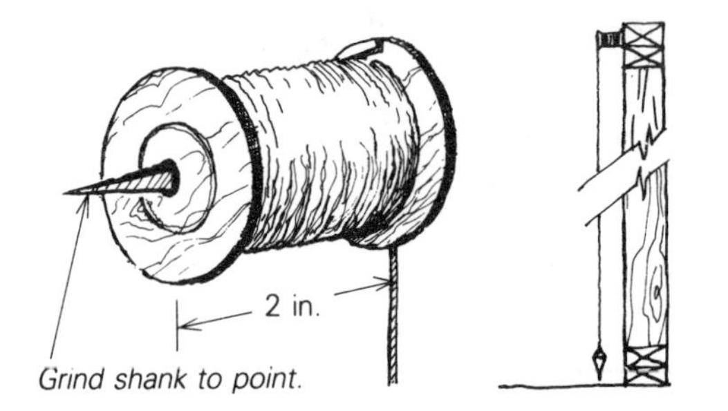

I keep my plumb-bob string wound around a spool exactly 2 in. long. I drive a screw into one end of the spool, cut its head off and grind the shank to a long, diamond-shaped point. Since I usually work by myself, I can drive the point into the top plate of the wall and dangle the plumb bob off the end of the spool. When the bob hangs 2 in. out from the bottom plate, I know the wall is plumb.

—Earl Roberts, Washington, D.C.

Homemade water level

Let's say you've just put four posts in the ground for a sun deck and now need to cut them all off at the same height. What do you do? If you have a builder's level or transit you're all set, but if you don't, this water level is a very simple and inexpensive way to solve the problem. It consists of any length of clear plastic tubing (or a hose with clear plastic extensions at the ends) and works on the principle that water seeks its own level.

First, fill the tubing with water, leaving a foot or two of air at the ends, and hold one end against the point to be transferred, as in the drawing above. Another worker takes the other end of the tubing to the first post to be cut off (post B in the drawing).

The person at post B holds the tubing against it while the person at the other end moves the tubing up or down until the water level in the tubing matches the level of the point to be transferred. When this has been accomplished, the waterline in the tubing at post B will be the same as at the transfer point.

Always transfer levels from the original point to lessen accumulated error. Remove all the air bubbles from the tubing—they can affect accuracy. Remember that no part of the tubing should be higher than the ends.

This tool is useful in many ways on a building site. We use it whenever we need to transfer a level point farther than the length of a carpenter's level.

—David Barker, Gardiner, Maine

Homemade scribe

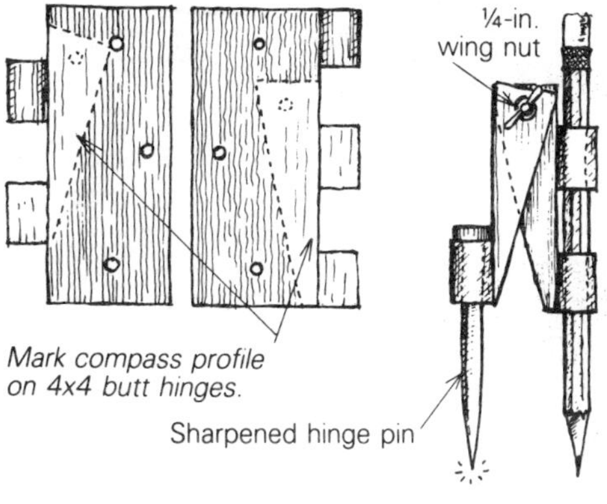

¼-in. wing nut

Mark compass profile on 4x4 butt hinges.

Sharpened hinge pin

The carpenter's scribes sold in lumberyards and hardware stores are so flimsy that they will self-destruct just rolling around with other tools in your nailbags. Sturdier versions are sold by mail order, but they cost around $20. Rather than spend the cash, I made the one shown here out of an old 4x4 butt hinge. It's accurate, nearly indestructible and easy to make.

I cut out the basic shape with a hacksaw, filed down the rough edges and drilled holes to accommodate a ¼-in. wing-nut pivot. A sharpened hinge pin makes a good turning point. The pin and the pencil are held in place by friction. If the slots are a little loose, a squeeze in the bench vise will tighten them up in a hurry.

—John H. Sandstrom, Fort Dodge, Iowa

Chalkbox conversion

Here's the setup that I use for my foundation lines. I take a new chalkline box and discard the line that comes with it. Then I fill its spool with braided nylon mason's line, and I put a loop on the loose end. Braided line costs more than twisted line, but it's stronger and snarl-free—an excellent line for straightening walls or checking foundations.

—Robert Francis, Napa, Calif.

Dual panel gauge

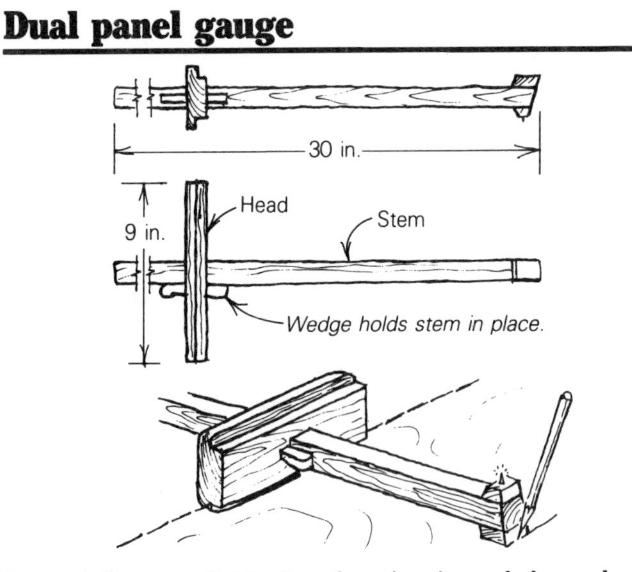

To mark lines parallel to the edge of a piece of plywood or other sheet material, I find a panel gauge is more accurate than measuring two positions and using a straightedge to draw a line through the marks. Many gauges are intended to scratch a line, but in some cases a pencil line is better. This gauge allows for both. The scale of this tool gives it a steady action and a reach of more than 24 in.—half a 4-ft. sheet.

The head of the gauge is a block of hardwood, rabbeted on each edge, with a hole cut squarely to take the stem. Hold the stem by a wedge driven into a slot in the head, as shown. The stem, made of a straight-grained hardwood, is built up at the working end to match the rabbets in the head and beveled for ease in scribing with a pencil. On the other side, make the scratch point by driving in a nail, cutting it off within ¼ in. of the wood and then sharpening it to a point.

—Percy W. Blandford, Stratford-upon-Avon, England

Crayon grip

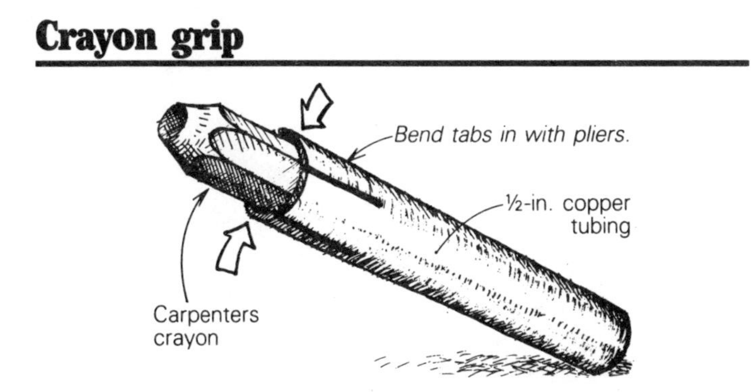

Here's a classy way to house a carpenter's crayon. First, get a piece of ½-in. copper tubing, about 3½ in. long. You can usually find one in the scrap pile after the plumbers have been on the job. With a hacksaw, make four slits about 1 in. long, an equal distance apart in one end of the tube. Bend the resulting tabs in slightly to taper the end of the tube. Now slide the crayon in from the other end, and you're ready to write. This holder (drawing, above) is less bulky and much cheaper than the ones sold at hardware stores.

—Mike Tsutsui, Stockton, Calif.

Caliper improvisation

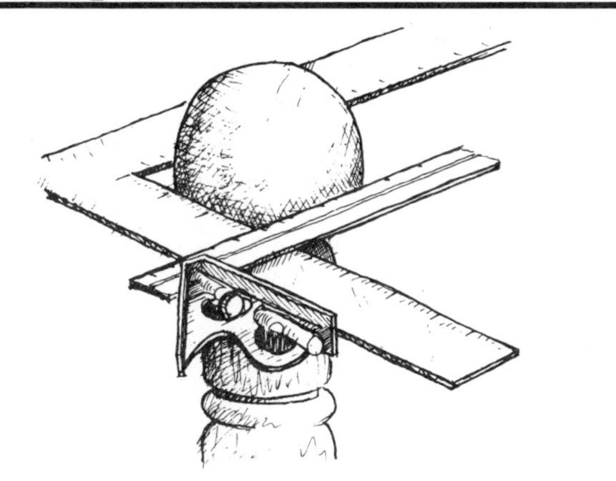

I was laying out a stair rail and needed to find the exact diameter of the ball atop a newel post. My calipers, naturally, were enjoying a day off back at the shop. I improvised by using my framing and combination squares as shown in the drawing above, and I read the diameter directly off the framing square. ***—Jim McConkey, Washington, D. C.***

Sawblade compass

As I was about to install a toilet on a recent plumbing job, I made a discovery. I needed to draw a circular cutline on the subfloor for the toilet flange, but I didn't have a compass handy. I did have my Sawzall, though. I removed its blade, and drove a nail through the hole in the blade into my center mark on the floor. With my pencil resting between two of the blade's teeth, as shown above, I was able to draw a perfect circle—just the size I needed.

—Anthony Revelli, New York, N. Y.

Snapping lines

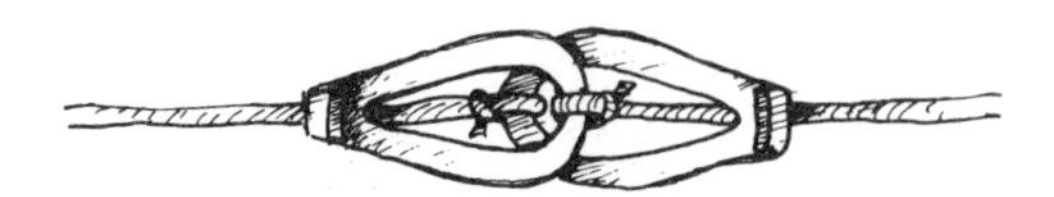

When snapping a series of chalk lines, as on roofs or siding, two people can hook their chalkbox lines at the clip by inserting one line through the other, as shown in the drawing. Worker A reels in the hooks to his end and they snap lines until worker B's line goes dry. Then worker B cranks his line while exposing A's for more snapping. There is no walking back and forth, and no time wasted rechalking the lines. ***—Jackson Clark, Lawrence, Kans.***

Lipstick on the job

More than once I've received a look of disbelief when I've asked a helper to get the lipstick from the truck. But the stuff comes in handy in a variety of situations.

When I need to mark the end of a door latch or deadbolt

strike, I rub some lipstick on the bolt, close the door and turn the bolt against the jamb to locate the proper spot to drill.

Drywall cutouts can be easily found by smearing the edges of the electrical box with the lipstick, hoisting the sheet of drywall into place, and pressing it against the outlet. Pull the drywall away from the box, and make your cutout on the lipstick marks for a snug fit every time. This principle works for paneling and siding too.

—Ernie Alé, Santa Ana Heights, Calif.

Stringline anchor

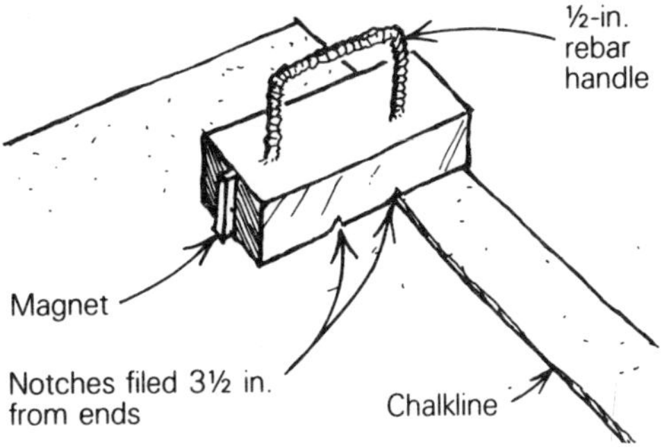

½-in. rebar handle

Magnet

Notches filed 3½ in. from ends

Chalkline

When it comes time to frame a house, I like to do the layout by myself. The solitude affords me time to work out the kinks in the plan without being barraged by questions from the crew. The problem is, it's tough to pop chalklines on a concrete slab (the typical foundation in my area) when you're working alone.

I solved the problem by getting a heavy piece of steel at a local scrapyard and welding a handle to it so I can lug it around more easily. This brick-size hunk of metal weighs about 20 lb., and it has no trouble anchoring the dumb end of a string line or steel tape. As shown in the drawing above, I affixed a bar magnet to one end of the anchor to secure the hook on a steel tape. Also, I filed notches along the edges of the anchor, 3½ in. from the ends. When the end of the anchor is flush with the edge of the slab, either notch can be used to anchor my chalkline for a 2x4 layout.

Because of its mass and sharp angles, the anchor sometimes gets used as a job-site anvil. It is also useful for tying up the dog when necessary.

—Don Huebner, Austin, Tex.

Eccentric layouts

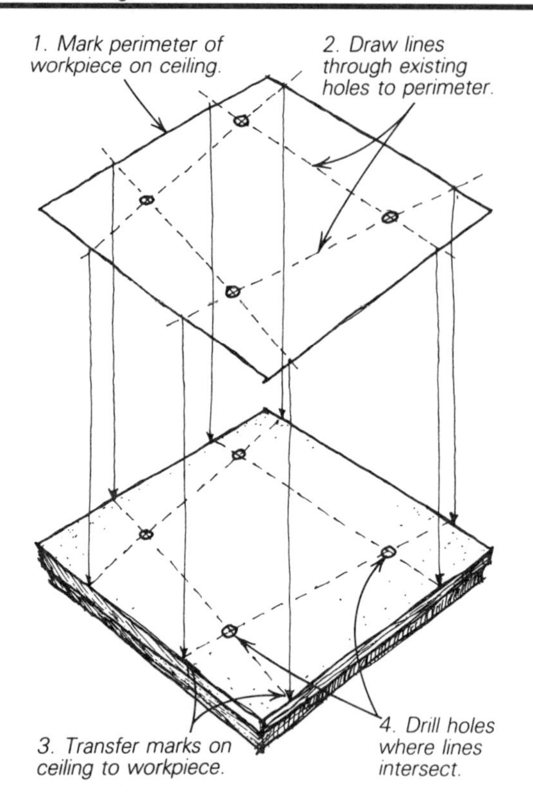

I needed to mount a wooden plate from which to hang a light fixture. I wanted to use four existing holes in the ceiling, which were by no means square to each other, to mount the plate. The problem was how to mark the hole locations on the plate for drilling.

My solution, shown in the drawing above, was to hold the plate in position on the ceiling and lightly draw its perimeter. Then I removed the plate and, using a straightedge, drew four lines passing through the existing holes. These lines extended just past the perimeter line. Next I held the plate against the ceiling and I transferred the marks onto the edge of the plate, as shown in the drawing. When I connected the marks on the plate with a straightedge, the intersections of the lines marked precisely the location of the holes.

—Barry Kline, Cleveland Heights, Ohio

Wind-up plumb bob

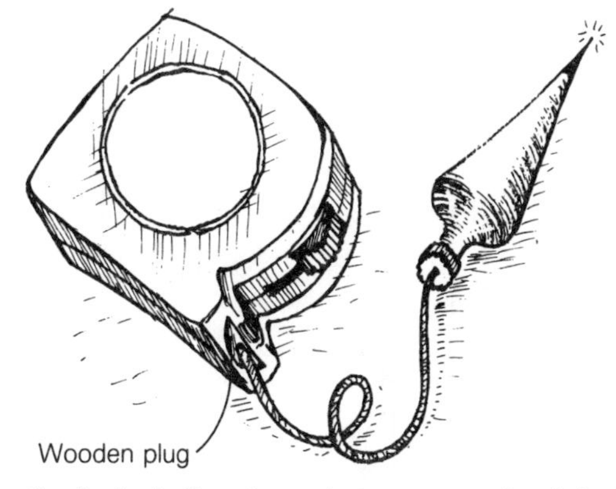

If you're tired of winding the string on your plumb bob, and you've got an empty spring-wound tape case lying around, try this. First, whittle a wooden plug to fill the tape slot, and drill a hole in it a little larger than the diameter of the string you use. Now remove the old tape and tie your line to the end of the spring. Let the spring and some of the line retract into the tape case, thread the string through the hole in the plug and press-fit the plug in place. Now tie a knot in the line or attach your plumb bob, and you're in business. You can get at least 25 ft. of line in a 25-ft. tape case, and with a larger-diameter line, the push lock will still work.

—Brent Lanier, Pleasant Gardens, N. C.

Stair-gauge straightedge

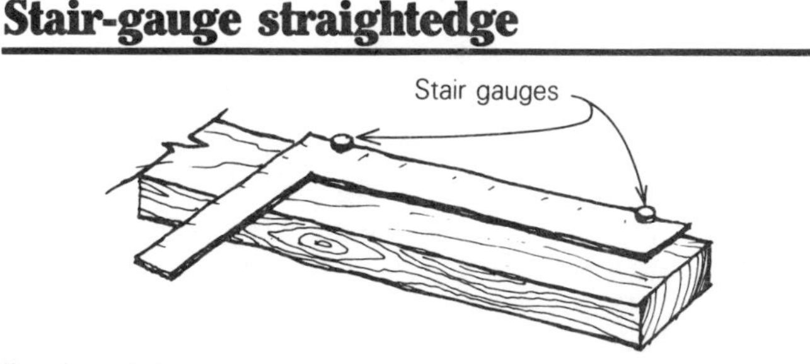

I've found that a framing square with stair gauges is a good tool for doing layout on wall plates. I clip the gauges on the long blade of the square, as shown in the drawing. The gauges easily align the blade along the edge of the plates, so that I don't have to worry about the blade getting cockeyed while I do my layouts.

—Will Milne, San Fransisco, Calif.

Leveling rod

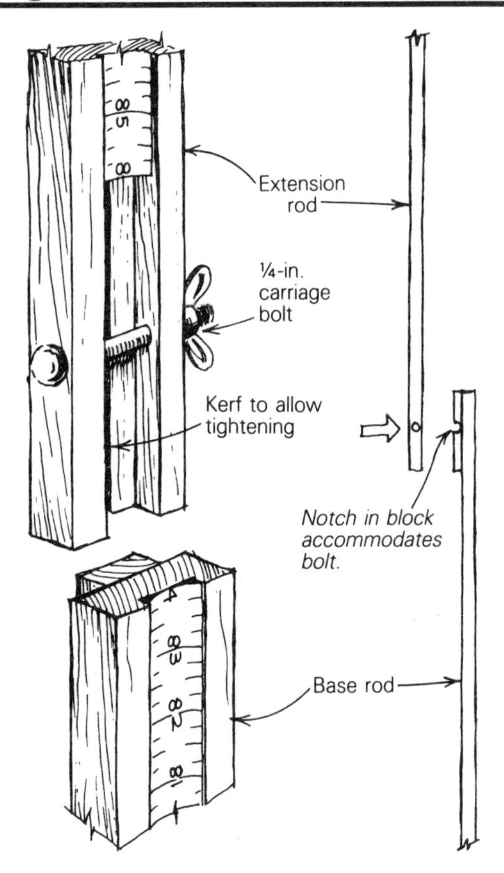

The drawing above illustrates an inexpensive design for an accurate, two-piece leveling rod made from a defunct 1-in. wide tape measure. To make the base rod, begin with a piece of straight, knot-free 1x stock, ripped to a full 2-in. width. I think 7 ft. is a good length for this section of the rod. Use a router or a dado head to cut a ⅞-in. wide by ¼-in. deep groove in the base rod. Use tin snips to cut a 7-ft. section of tape, and press-fit the tape into the groove. It should stay put.

To make an extension rod, cut another groove in a similar piece of wood, an inch or so shy of 7 ft. long. At one end, deepen the groove to ½ in. for the first 5 in. of the rod and cut a saw kerf 5 in. long in the center of the groove. Next, drill a hole for a ¼-in. carriage bolt, as shown in the drawing. Now you can press the continuing portion of the tape into the groove, starting at the end of the kerf.

At the top of the base rod, glue a ⅞-in. wide block, about 5 in. long and ½ in. thick. This block has to be notched to accommodate the bolt in the extension rod. To use the extension, press the block at the top of the base rod into the groove in the extension rod. Adjust the extension up or down until its tape is even with the top of the base rod, and tighten the wing nut.

When the extension rod is not in use, I turn it upside-down and reattach it to the base rod, out of the way. This rod is accurate, and I think it's easier to read than commercial ones, which cost $60 or more. ***—Jim Reitz, Towson, Md.***

Erasing chalklines

I had just snapped a chalkline across the face of some very expensive western red cedar paneling when I realized I had measured incorrectly. To lift the chalk out of the very porous veneer without smearing or rubbing it in, I mixed up a glob of stiff flour-and-water dough and just dabbed at the chalk. Followed up with light sanding, this made the goof virtually disappear. ***—James Bolker, Lyle, Wash.***

Documenting projects

I have a simple, accurate way to document the placement of the framing, wiring and plumbing parts that go into a construction project: photographs. If I see the possibility of future remodeling, or special use for a particular wall (such as carrying a row of cabinets), I photograph the wall before the drywall installation. If possible, I stretch a tape measure in a corner of the photo to give me a scale. When excavation is part of the job, I take pictures of the water lines, septic systems and other underground utilities.

Over the years I have saved many hours and avoided a lot of mistakes by referring to my photo file. When I'm done with a job I leave the photos with the owners in a package that includes the plans of their projects. I retain the negatives in my file. ***—Howard Furst, Bellingham, Wash.***

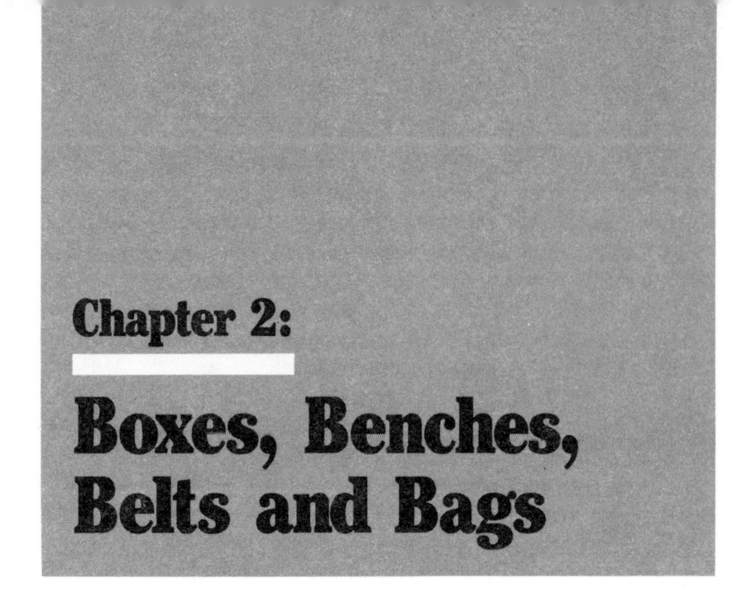

Chapter 2:

Boxes, Benches, Belts and Bags

Modified adjustable tool belt

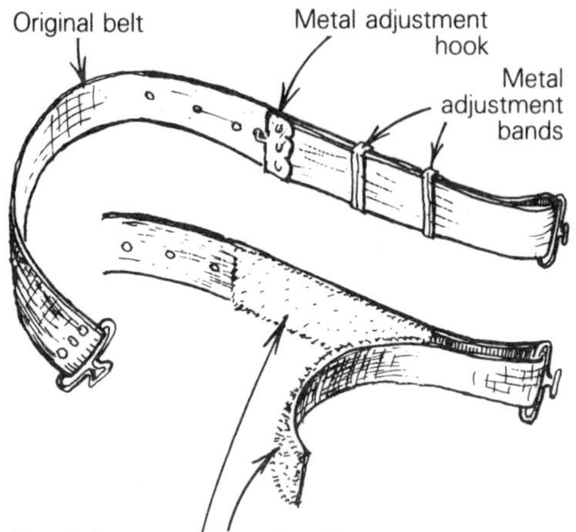

Touch-fastener strips allow fine adjusments in belt diameter.

If you are tired of your tool belt coming unhooked every time you pick it up, and are fed up with the metal adjustment bands digging into your gut, try the following alteration. Cut off the metal adjustment hook (preferably with a hot-melt cutter to seal the end of the belt). Now remove the metal adjustment bands and throw them away.

To modify the belt, glue a strip of 2-in. wide touch fastener about 1 ft. long to the inside of the belt where the overlap occurs. I used several coats of contact cement for a flexible and tenacious bond. Make sure to put the hook side of the material so that it faces away from your skin. This will make the belt more comfortable if you work in shorts during the summer.

Refined in this manner, your belt can be quickly and accurately adjusted. In my experience, the fastener has never come undone. *—Alan Carrier, Carbondale, Ill.*

Nail-pouch chisel roll

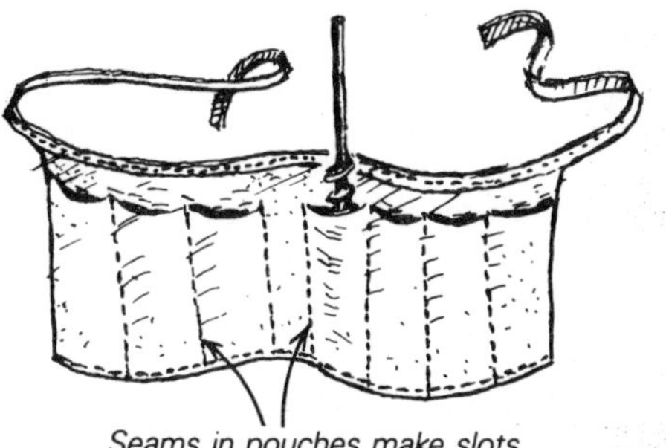

Seams in pouches make slots for bits or chisels.

Many lumberyards give away canvas nail aprons. While the aprons may be flimsy and of little value as nail holders, they can be easily modified into a rollup tool pouch for auger bits or chisels. As shown above, sew some seams into the pouches, and you've got slots for some of your edge tools.
—Samuel D. Jannarone, Upper Nyack, N. Y.

Tool-belt tip

I use separate pouches on my tool belt depending on whether I'm doing rough carpentry, drywall or electrical work, and I finally found the perfect belt to hold them—a skindiver's weight belt. These belts are made of nylon webbing, so they are plenty tough. Even so, my belt is comfortable, even when I'm bending or stretching, because I can adjust it up or down on my waist and hips with the quick-release buckle. *—Michael Sweem, Downey, Calif.*

Sliding tool chest

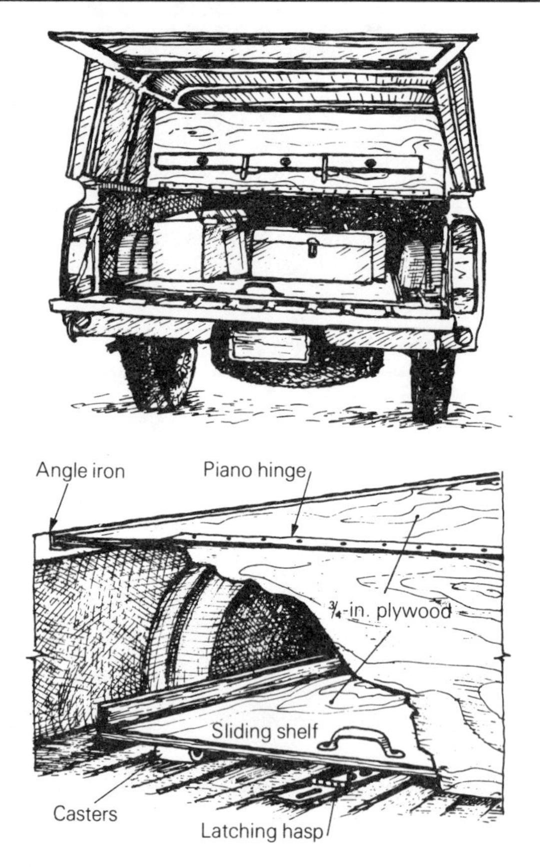

I was able to solve the problem of keeping my tools locked up and yet accessible in my pickup camper shell with this sliding-shelf tool locker.

I mounted a piece of angle iron along each side of the bed, level with the top of the tailgate. Then I cut a ¾-in. plywood shelf about 4 ft. long and as wide as the bed, and bolted it to the angle. I hinged another piece of plywood at the back of this where it would fold down just behind the wheel wells. I added a latching hasp and now had a lockable compartment for my tools as well as a storage space above them. For inside this compartment I cut ¾-in. plywood narrow enough to slide between the wheel wells. I edged the plywood on three sides to keep toolboxes from sliding off, put a handle on the front, and attached casters to the bottom. This gave me a drawer that could be locked in place when the lid was closed. With the tailgate open, this sliding shelf will reach far enough to give access to everything on it.

Long material can be carried out the back of the camper shell supported by the top of this locker and the tailgate, which are at the same height. The entire system can be removed in minutes if the full use of the bed is needed. Also, I have discovered that the inside of the lid is a good place to store such awkward tools as my 4-ft. level.

—Kevin Ireton, Dayton, Ohio

Blue-jean nail bags

When my blue jeans gave out at the knees, I recycled them into a carpenter's apron. To do this, take an old pair of jeans, with good back pockets, and go after them with a pair of heavy-duty fabric-cutting scissors. Cut away the legs about an inch below the pockets, and then up the side seams to the belt line. Now continue the cut under the belt line, leaving the rear half of the jeans and a strap to put around your waist, as shown in the drawing above.

In addition to being free, this apron is guaranteed to fit (assuming you've recycled your own jeans) and the button is already in place.

—Ray Jorgenson, Center Ossipee, N. H.

Offset handle

I used to carry a standard carpenter's wooden toolbox, with room for hand tools in the bottom, a nail tray on top and a wood clothes-pole handle. But I found that when I carried it, it would bump into my leg if I didn't hold it away from my body. This was hard on my arms and legs, so I offset the handle, bringing it an inch closer to my side. Now the contents don't shift to the outside, and the box is a lot easier for me to carry.

—Steve Hunter, Berkeley, Calif.

Quick clamp support

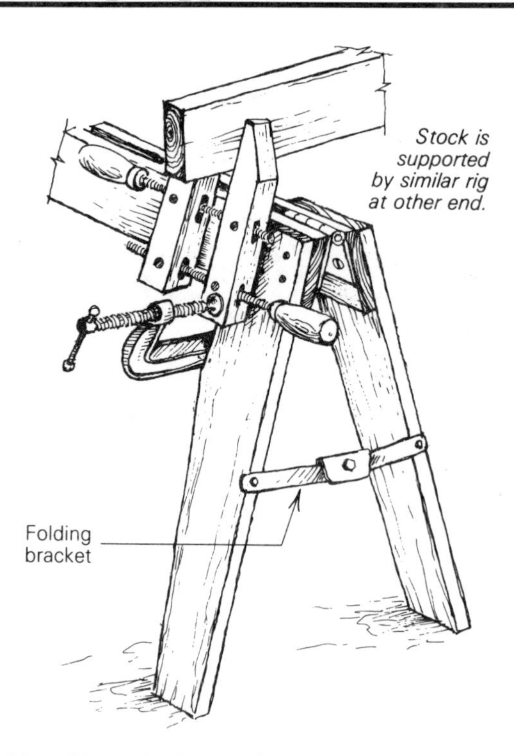

One problem I have had in working on site is not having a workbench to hold stock for edge-planing and similar operations. I solved this by using a handscrew clamp and a C-clamp in combination, as shown in the drawing above. The C-clamp secures one foot of the handscrew clamp to the leg of the sawhorse, leaving the other foot free for adjustment. This setup works well on conventional sawhorses and on some folding ones. The only alteration I'd recommend on folding sawhorses is to use card-table brackets on each side of the horse to keep the legs from folding or splaying during use.

—Ben Erickson, Eutaw, Ala.

Chop-saw box

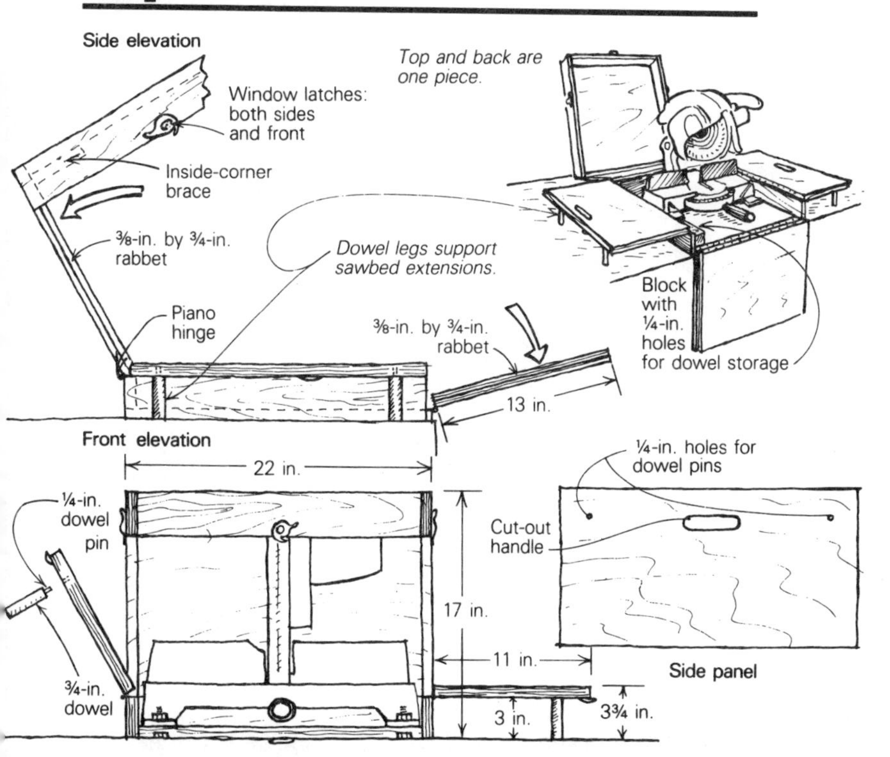

The motorized miter saw is a great tool, but it does have some drawbacks—its protrusions make it awkward to carry and its bed is often too short to accommodate the work at hand. I finally built a carrying case for my Makita that solves both problems for me.

My carrying case is shown in the drawing above. It's made of ¾-in. birch plywood, and it measures 22 in. square by 17 in. high. Its sides, front and back are all attached to the base with piano hinges. When folded away from the saw, the back becomes a dust deflector and the sides turn into sawbed extensions. The extensions rest on ¾-in. dowel legs, which store inside the box for transport. The ¼-in. dowel pin recessed into each leg fits snugly into a ¼-in. hole drilled in the exterior faces of the side panels when they are being used as table extensions.

When the sides are up and the lid is in place, the box is held together by window latches on the sides and front. Altogether, the box weighs approximately 20 lb. I gladly carry the extra weight, knowing that I will have a good work surface wherever I go. ***—Ron Austin, Park Ridge, Ill.***

Tool sling

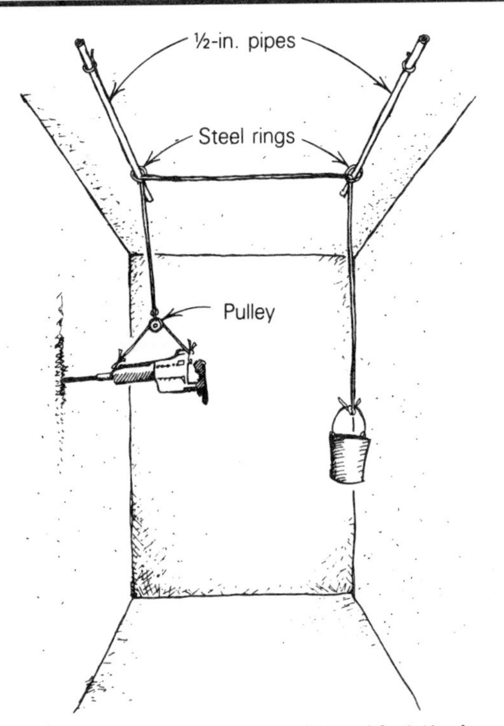

Using a sling to support a heavy tool in mid-air isn't a new idea, but I think the system I use is an improvement over other methods I've seen. Instead of suspending the tool from a single hook, I hang it from a steel ring that slides along a piece of ½-in. pipe. This allows me to move from side to side with the tool without the restraint of a fixed mount.

I usually use this technique when I'm working on a vertical surface with my Bosch demolition hammer. The hammer is on the working end of the rope, and a bucket containing sand is on the counterweight end. In the setup shown in the drawing, I had to remove the stucco from the walls of an old porch. On this job, I used two pipe rails mounted on the ceiling with open-eye hooks. The double-pipe system was especially nice in this case because it kept the counterweight away from the work, and allowed me to move easily from wall to wall.

—Peter Fenerin, Palo Alto, Calif.

Belt-buckle upgrade

My carpenter friends and I have all agreed for some time that the weak link on a tool belt is the clasp. Some twist-grip buckles are flimsy, and they can all be clumsy to disengage, especially if you have to remove your belt in a precarious spot.

As a remedy, I removed my old clasp and replaced it with a seat-belt buckle. I tried both the push-button type and the spring type, and decided I like the spring type better. Now I can easily remove the belt when I want to, and with one hand at that. ***—Evan Disinger, Lemon Grove, Calif.***

Lightweight workbench

I got the idea for the workbench shown above from a couple of old West Virginia carpenters. I've been using this design for years now, and the benches are unbeatable for holding work that's being drilled, cut or nailed. Whenever I do drywall work, I lay two benches on their sides and cover them with planks and plywood to make a low scaffold. It's just the right height for working on an 8-ft. ceiling.

I've found these dimensions to be right for me, but of course they could easily be altered to suit the individual. The material for each bench costs less than $10. Nail them together with 8d cement-coated nails, and you're ready to work. ***—Howard Goldblatt, McLean, Va.***

Folding workbench

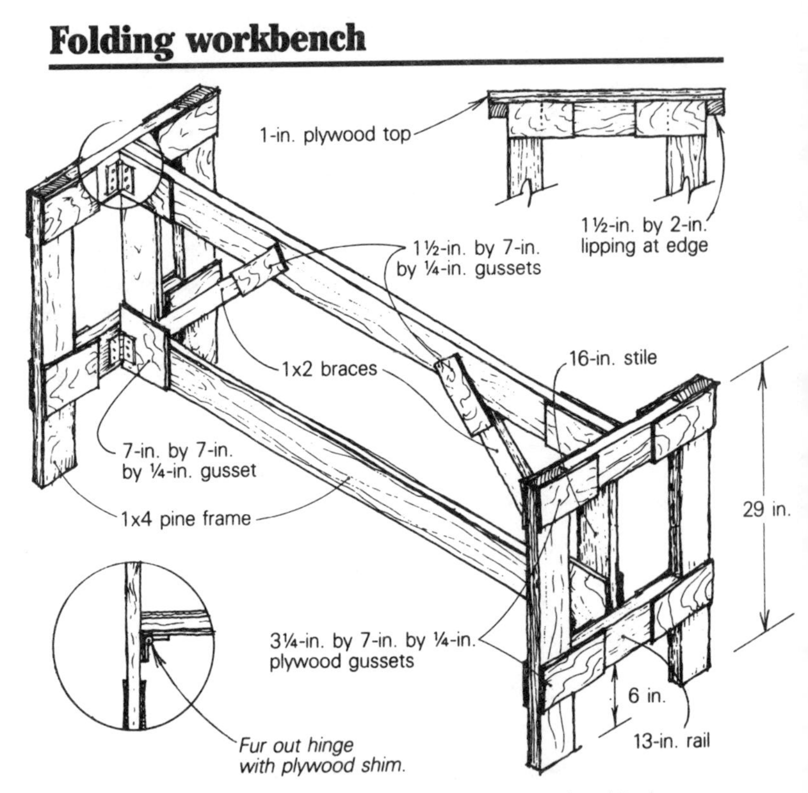

When I left my amateur standing for an apprenticeship in carpentry, I also left my overbuilt workbench in the basement. One evening, when I was glancing over plans for site-built roof trusses, I realized that the same principles could be used to design a solid, easy-to-carry bench. Glued-on gussets are the secret. They reinforce a lightweight frame where it's needed most—at the joints. For my bench's frame I used 1x4 clear pine. At first I had my doubts about the strength of ¾-in. stock, but I'm glad now that I didn't use 2x4s.

I started by cutting four 29-in. legs, knowing that I'd probably shorten them. Then I cut four 13-in. rails, as shown in the drawing above. This gives a finished width of 20 in. for the end frames. This might seem too narrow, but the bench is very stable in use. The top rails join flush to the top of the legs, and the bottom rails are 6 in. above floor level. This allows plenty of length for trimming the bench to the correct height.

I fastened the end frames together temporarily with corrugated fasteners, coated the gussets with glue and stapled them across the joints. Staples are fine for this; the strength comes from the glue.

I made the mid-frame 5 ft. long, but if I had to do it over I'd up it to 5½ ft. The bench would be a little harder to maneuver through tight places, but the extra stability would be worth it. The mid-frame stiles are 16 in., to match the dimension between the rails of the end frames. To resist racking, I added a 1x2 brace at each corner of the mid-frame.

To join the three frames, I used pairs of old 2½-in., tight-pin butt hinges—a pair at each end—so that the whole thing folds flat for transport or storage. The hinge leaves on the end frames need to be furred out with plywood the same thickness as the gusset.

My bench top is made from a 4x8 sheet of ½-in. CDX plywood, cut in half lengthwise and laminated together with yellow glue. I glued the two crowned sides together to ensure that the finished piece stays flat.

Once the top was dry, I glued 1½-in. by 2-in. lipping on each long edge. This strengthens the plywood, holds the top in place and provides a handy purchase for clamps. Register blocks at the end of the top keep it centered on the base.

I've finally gotten its height to the right point for me—I'm 5 ft. 10 in. tall, and the bench is now 27½ in. I can plane comfortably on it, get a knee up to crosscut and still avoid the backache of working on low horses.

—T. D. Culver, Cleveland Heights, Ohio

Custom roof rack

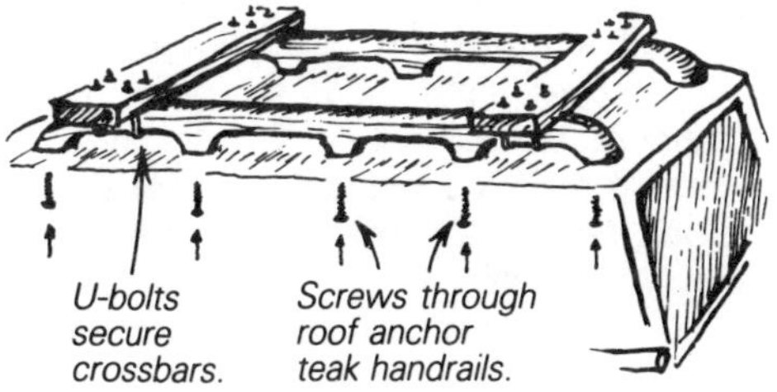

I have a sturdy fiberglass shell on the bed of my pickup truck. While it's handy for storing tools and many supplies, I needed a rack on top to carry oversize cargo. I come from a nautical background, so I wanted my new rack to reflect my long-standing interest in finely crafted boats. This led me to the local boat-supply shop, where I found a good selection of teak hand-rails. They come in various lengths, depending on the number of loops in the handrail. Each loop is about 14 in. long. The rails I selected are the four-loop variety.

I attached the handrails to the roof by running screws with ⅝-in. washers through the fiberglass roof from inside the shell. The crossbars are pressure-treated 2x6s, secured to the handrails with a pair of U-bolts at each connection. Now I've got a sturdy, versatile roof rack that looks sharp, with lots of places to anchor a line.

—Chuck Keller, Marblehead, Mass.

Magnetic nail pouch

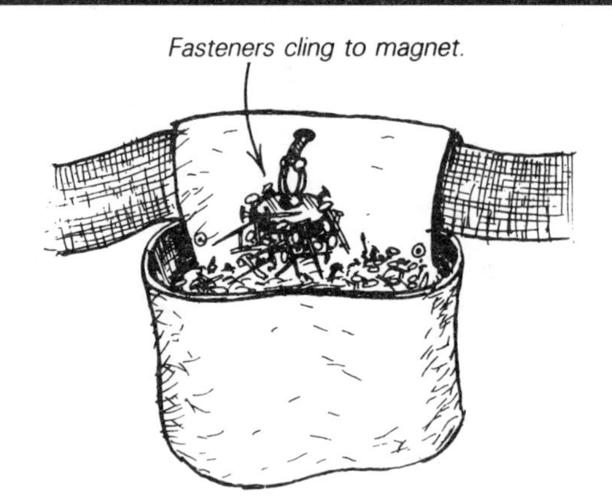

On a large drywall project you can wear out a pair of gloves (or your fingertips) just by picking screws or nails out of your nail pouch. To prevent this problem, I've attached a magnet to my tool belt with a strong rubber band. I placed the magnet over the center of my nail pouch, as shown in the drawing above. Located here, the magnet can be extended down into the pouch, where it picks up a supply of fasteners. Then I can quickly pluck them off the magnet without jamming the tip of one under my fingernail.

—Bob VonDrachek, Missoula, Mont.

Panel hook

A 4x8 sheet of drywall, pegboard or plywood can be a nuisance to carry. With my device, one person can easily maneuver these bulky panels. I fabricated the metal hook shown in the drawing on the facing page from a scrap piece of 12-ga. cold-rolled steel. A short length of rope is threaded

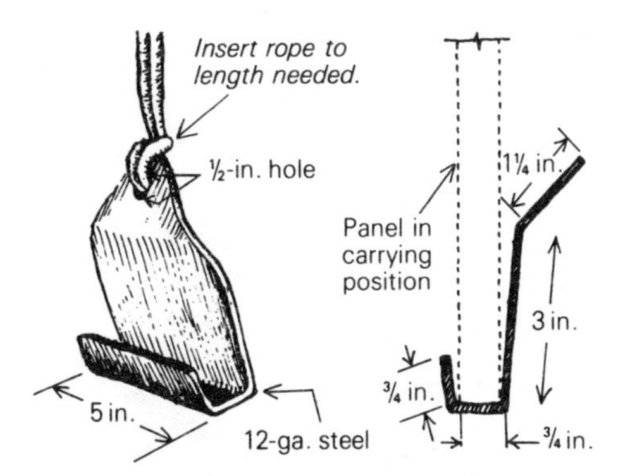

through the ½-in. diameter hole and serves as an adjustable handle. By hitching up on the rope slightly and adjusting the panel's center of gravity I can negotiate stairs, up or down, with no problem. ***—George Eckhart, Kenosha, Wis.***

Nailbag liners

I save half-gallon milk cartons, cut them in half, and use them for nail containers. Their 4-in. by 4-in. size fits nicely into the large bag on my nail pouch. When I'm done with one kind of nail, I remove the carton from the pouch and return it to its storage cabinet. This practice keeps the number of miscellaneous nails in my pouch from getting out of hand. ***—Sam Francis, Bozeman, Mont.***

Plywood carrier

I recently had to move a stack of ¾-in. plywood without a helper. After a bit of head-scratching, I worked out a system using a short wrecking bar with a 90° claw on one end. To use the bar, I lift one end of the plywood slightly and swing the claw end of the wrecking bar under the panel. Then I shorten my grip on the wrecking bar, pick up the sheet and use my other arm to steady the load. It's not much different from the usual design for a panel carrier, but my wrecking bar is always on site with me.

—Brad R. Johnson, Chicago, Ill.

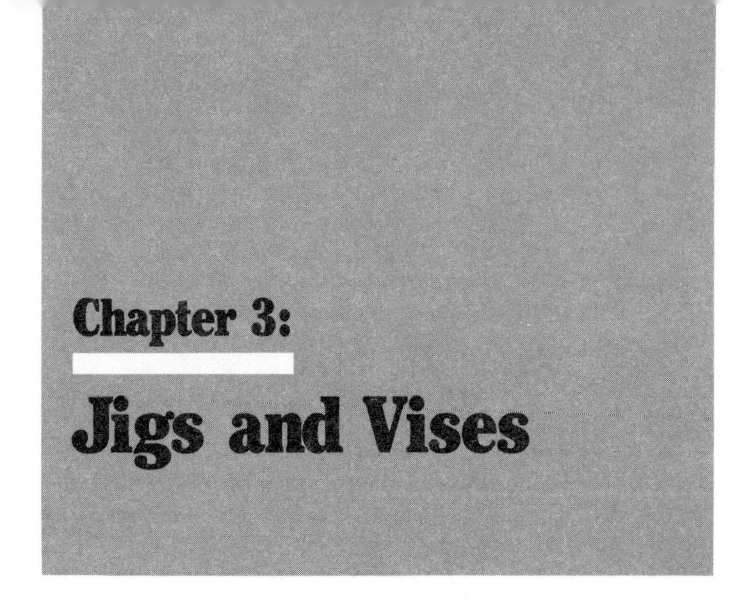

Chapter 3: Jigs and Vises

Cutoff fixture

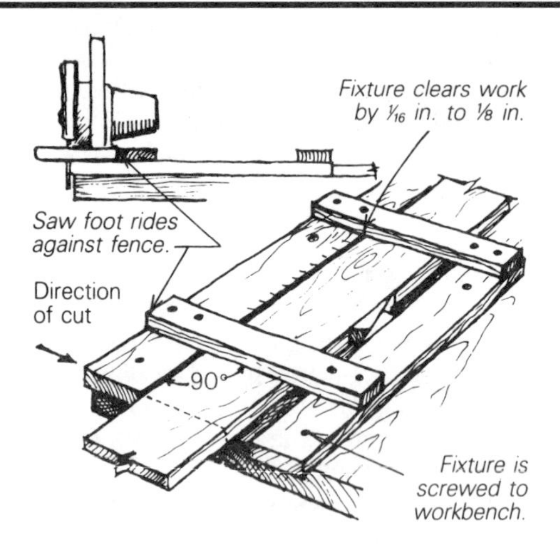

I made this bench-mounted cutoff fixture for a circular saw out of scrap materials in less than an hour, and it saved me the price of a saw protractor and a trip into town. It's simply a pair of parallel rails, held in place by two crossbars. One of the crossbars doubles as a fence to guide the saw's foot. I made the one shown in the drawing for 90° cuts, but it's easy to nail the fence at some other angle.

The distance between the rails depends on the width of the stock being cut. I positioned them so I could use two wedges to keep the stock from slipping, giving me a free hand to catch the cutoffs. I made reference marks on the near rail so that I could mark the length needed for the next cut without using a tape measure each time.

—G.R. Livingston, Firestone, Colo.

Miter jigs

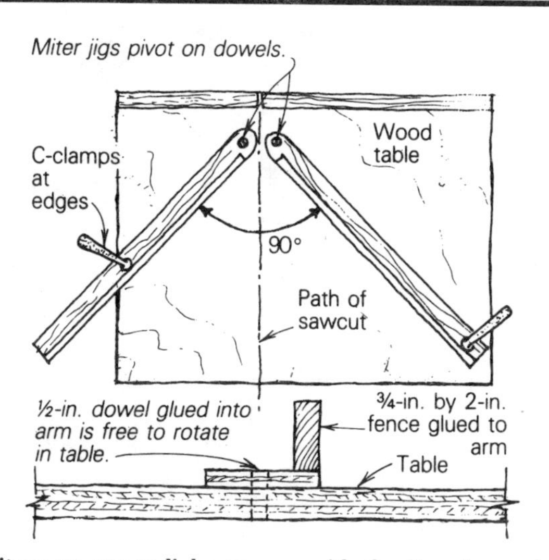

I cut miters on my radial-arm saw with the jigs shown in the drawing. In the plan view, the path of the sawblade should bisect the miter fences—a 45° cut on both sides. But as long as you are cutting stock of the same width using both fences for each joint, and the fences are exactly 90° apart, the miter will fit perfectly even if the saw is slightly off. To make sure of the angle, I hold my framing square between the two arms when I clamp them down.

The arms can be plywood or clear stock—just something that will remain true. I screwed and glued ¾-in. by 2-in. fences to the arms after I made sure that the saw motor would clear them. The pivots are short sections of ½-in. dowels, glued into the arms. ***—James Baldwin, Chester, N. J.***

Another cutoff fixture

I seldom use radial arm saws when I'm working on a job. They are expensive, difficult to move around and constantly out of adjustment. The alternative is to use the simple cutoff fixture shown in the drawing above, a circular saw with a shoe that is parallel to the blade, and a $15 saw protractor.

The fixture can be made from 2x lumber of any width, and should be long enough to support the boards you are cutting. Attach a block across the face of the fixture at one end. Make sure that this end is perpendicular to the edge of the jig. This block should be back-beveled along the end grain on the working side of the jig to prevent sawdust buildup when it's in use. A 2x4 block can then be nailed to the side of the fixture, with the bottoms flush. This is the protractor stopblock. It should be long enough, with the additions of the protractor arm length and the width of your saw shoe, to produce the cutoff length that you want.

Butt the piece to be cut against the end stop, set the protractor across the board and against its stop, and run your circular saw against the leg of the protractor. This method will produce fast, clean, accurate and repeatable cuts of any angle for everything from joist blocking to finish work without measuring and squaring each board.

If you are production-cutting several lengths, nail protractor stops in several locations along the length of the fixture. Clamping the stop is faster if only one or two cuts are needed at that length. *—Ron Davis, Novato, Calif.*

On-site vise

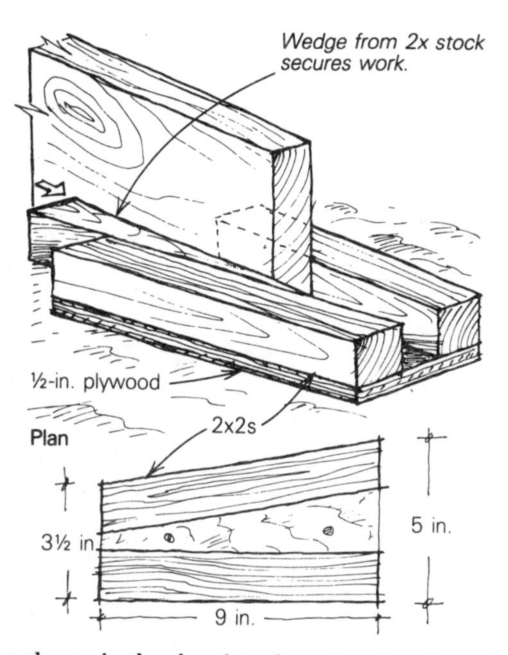

The vise shown in the drawing above is a handy and inexpensive tool for securing work up to 1¾ in. thick. It is made of 2x2s screwed and glued to a piece of ½-in. plywood. On the job site, I usually screw one to a subfloor, deck, bench or a sawhorse. The workpiece fits into the tapered slot between the 2x2s, and it's held fast by a 2x wedge driven into the gap. To level the workpiece, I shim its other end with a piece of ½-in. plywood. This rig works especially well for planing the edges of boards.
—Carl Meinzinger, Guemes Island, Wash.

Post cutoff jig

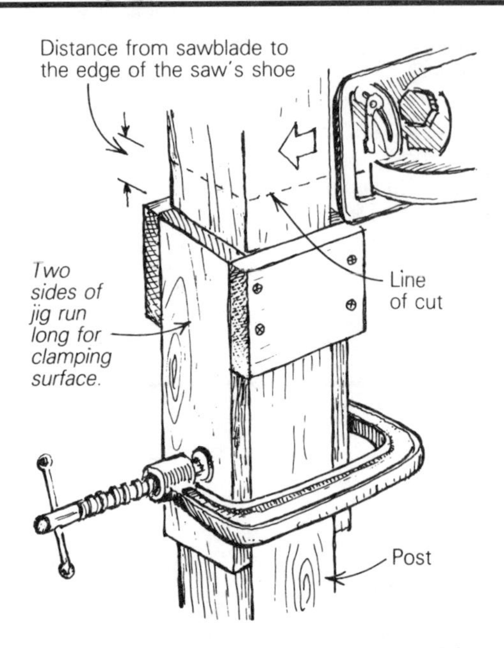

This year we have been doing a lot of decks and fences, and to simplify leveling the tops of all the posts I devised the jig that is shown in the drawing above.

The post-cutoff jig is a four-sided box with inside dimensions 1/16 in. larger each way than the cross section of the posts. The four sides of the box meet in the same plane at the top, where they are secured with screws. The two long, narrow sides hang down about 10 in. for a clamping surface.

First we plumb and stabilize all the posts, letting their tops run wild. Then we mark the desired height on one post, and transfer that height to the rest using a water level. Below each mark we measure down and scribe a second mark. This measurement is equal to the distance from the sawblade to the edge of the saw's shoe.

Now we drop the jig over the post, clamp it so the top edge is on the lower mark and use the top of the jig to guide the saw along all four sides of the post. If the piece that's being cut off is longer than 1 ft. or so, then have a helper lift up on it as you're finishing the cut so that it won't bind on the blade or fall on you.

—Timothy Pelton, Fairfield, Iowa

Folding horses

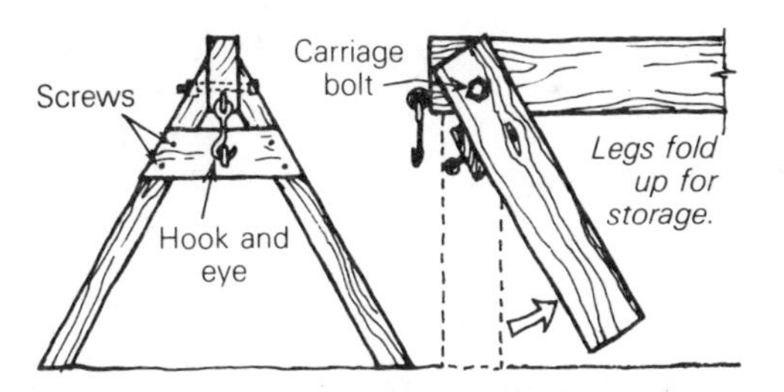

I can transport my wooden sawhorses in the trunk of my car and store them in tight spaces because I've designed them with folding legs. The trick is to build each leg pair as a unit with the cross brace screwed in place. Attaching each leg unit to the horizontal member with a single carriage bolt lets the legs pivot. Ordinary hook-and-eye catches on each end will keep the horses steady while they are being used.

—Barry Bower, Baddeck, N.S.

Taper jig

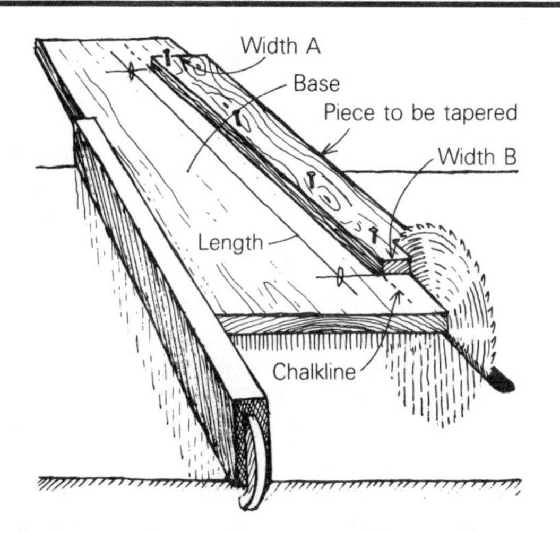

Here's a jig for tapering a piece on a table saw. For the base, take a straight 1x10 or 1x12 and measure the length of the taper, squaring off the end marks. On these lines, mark the width of the taper (A and B in the drawing) at each end by measuring in from one edge of the base. Connect these points with a chalkline. Tack the piece to be ripped to the base along the chalkline, and adjust the saw-fence to the width of the base. Set the blade deep enough to cut both pieces, and you're ready to run it through, as shown.

If the piece you are tapering is finish trim and you're concerned about the nail holes showing, calculate where the nailing will come on the piece and use these same points for tacking the trim down to the 1x base for the saw run.

—Sam Yoder, Cambridge. Mass.

Power miter-box jig

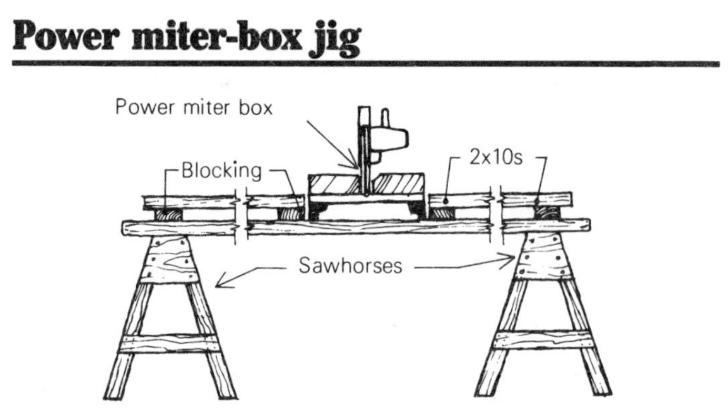

On a construction site, a convenient setup for a power miter box saw is essential if the tool is to earn its keep. A good setup must combine easy removal of the saw for secure lock-up, long supports on each side of the saw, easy storage, and a waist-high saw table. My jig, shown in the drawing above, allows for lift-off removal of the saw, which means it can be brought to a contruction site before the building is enclosed and can be taken away each night.

Because accurate cuts are so easy to make with one of these saws, inexperienced workers can get excellent results for such tasks as rough framing. If the jig has sufficiently long side boards, a stop for identical-length studs may be fastened to them. I haven't found it necessary to fasten my saw to this jig, but if you want you could insert dowels through the bolt holes in the box into the 2x10.

My 10-in. saw will not quite cut through 2x6 material unless I use one of these two tricks: On 90° cuts, I first bring the saw down through the wood as far as it can go, leaving the last ½ in. or so uncut. Then I lift up the front edge of the board to complete the cut. If I am making numerous cuts, or cuts at an angle other than 90°, I fasten a suitable length of ½-in. plywood along the side boards and across the saw table in order to jack the bottom of the lumber up into the saw's range. ***—Philip Zimmerman, Berkeley, Calif.***

Lumber roller

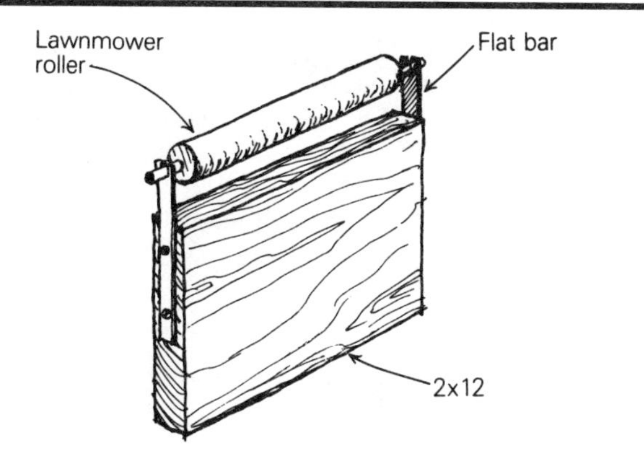

I use the device shown in the drawing above as an off-feed support for running long stock through my table saw. It consists of an old lawnmower roller attached to a piece of 2x12 with two notched pieces of steel flat stock. I clamp the support in my Workmate, where I can adjust it easily for height. ***—Patrick Lawson, Sooke, B. C.***

Another sawhorse

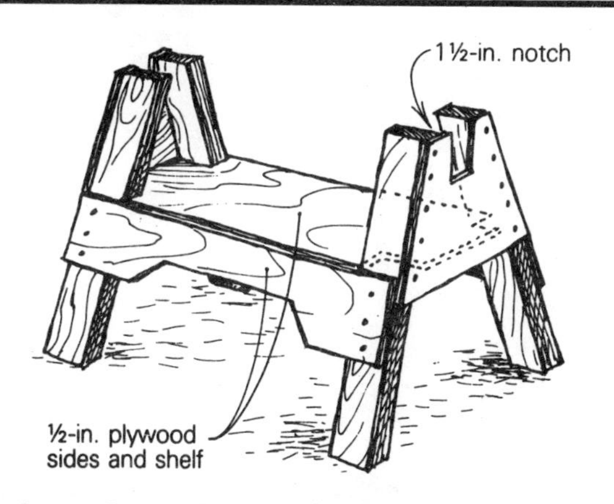

I made the sawhorse shown in the drawing out of scraps, and I like it better than any I've used in the past. Its main feature is a 1½-in. notch at each end, which accepts a 2x crosspiece of any width. This allows me to replace this member easily when it gets abused. I can put in the exact height I need to match the level of a work table or bench. The notches also hold narrow pieces when I'm working on an edge. Two-by stock drops right in, and 1x material can be wedged tightly with shims.

To make the horse, I started with two ¾-in. plywood gussets on the ends screwed and glued to 2x4 legs. Then I let in a ½-in. plywood shelf notched around the legs. The shelf bears on the ½-in. plywood side pieces, which go on last. By making one horse a little longer than its brother, I can tuck the shorter one into it for easier transport.

—Sam Yoder, Cambridge, Mass.

Low-budget outfeed support

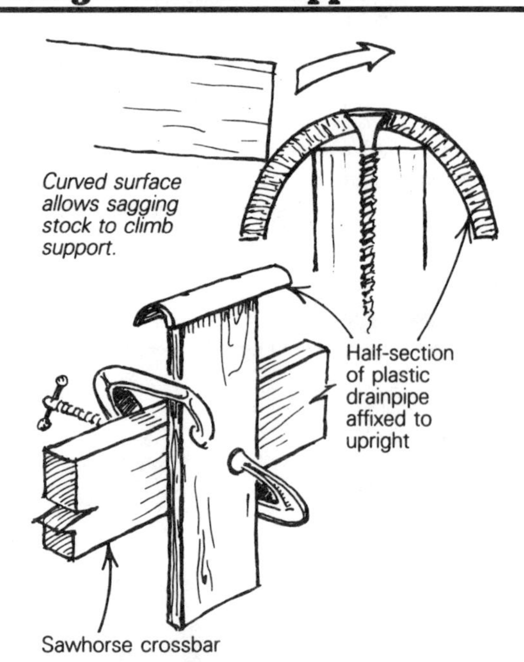

Roller supports are handy items to have around the workshop or the job site, and like C-clamps you can never have too many of them. Unfortunately, they are not cheap. To make some low-budget versions, I use plastic drainpipe offcuts screwed to wooden uprights. Then I clamp them to sawhorse crossbars at the desired height.

As shown in the drawing, I use a piece of pipe that has been cut in half lengthwise for the sliding surface of the support. For 3-in. pipe, a 2x upright will do. For smaller pipe, use 1x stock. Affix the pipe to the upright with countersunk flathead screws. In use, the plastic pipe offers very little resistance to a piece of stock sliding across it, and it will even allow a piece of stock with some sag in it to climb the curved surface.

—Bob VonDrachek, Missoula, Mont.

Knock-down horses

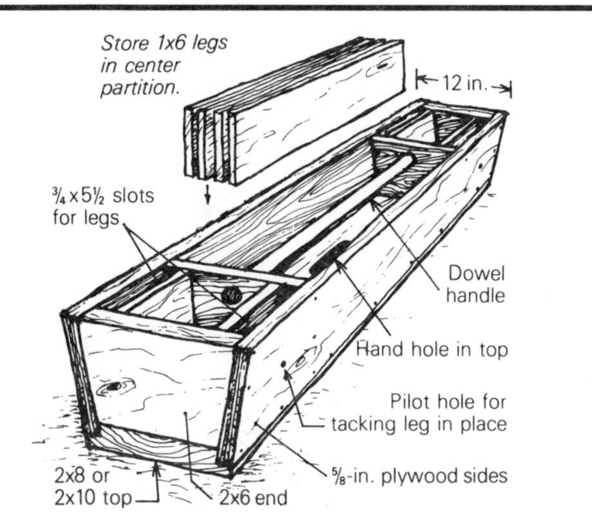

I like the convenience of knock-down sawhorses. The commercial ones are generally too flimsy, and their small legs sink into the mud after you've got them loaded. After a lot of experimentation, here's how I build a strong, compact stackable knock-down sawhorse that doubles as a toolbox (drawing, above).

I use a 2x8 or 2x10 for the top. This gives me a bench surface that I can work on or set my miter box on. I cut a hand hole in the top for moving the horse around when it's set up.

The end caps, central dividers and interior leg supports are 2x6s. The dividers hold a dowel that serves as a handle when the horse is disassembled and inverted. The compartment that is formed by these dividers should be the length of the legs so that they can be stored there, along with a few tools that you want to bring along.

The slots for the legs are ¾ in. by 5½ in., a snug fit for the 1x6 legs. I drill a pilot hole through the legs and the ⅝-in. plywood sides so that I can use a 16d nail to hold the legs to the body of the sawhorse when it's in use.

I cut different-length legs for different tasks, and I build different-width horses as well so that they stack easily. Aside from being slightly heavy, there's only one drawback to these sawhorses—you can't leave them on the job.

—George M. Payne, Olney, Md.

Stepladder standoff

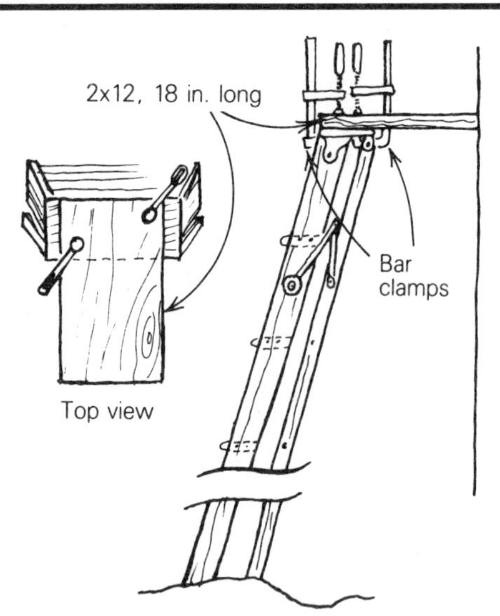

Sometimes when I work on windows or exterior trim I find myself lacking the appropriate ladder. In accordance with Murphy's law, an extension ladder is too long, and a stepladder propped against the wall ends up too close to the job to work comfortably. Under these conditions, I use a sturdy 2x12 standoff, 18 in. long.

I use a pair of short bar clamps to secure the 2x12 to the top shelf of the ladder, as shown in the drawing above. The standoff transfers the force of the ladder to the wall, and gives me room to maneuver around the task. As a bonus, the standoff makes a handy shelf for tools and materials.

—David Nauman, Marlboro, Vt.

Outboard-roller stand

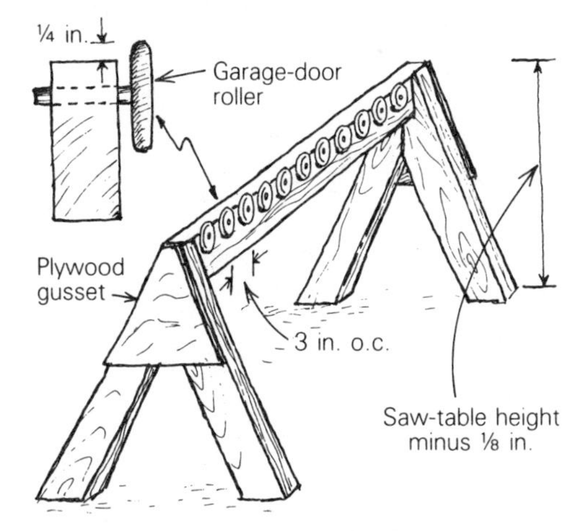

One of the most awkward operations to perform in my tiny one-man cabinet shop is crosscutting a full sheet of plywood. The outboard roller stand shown in the drawing above has made this task a lot easier for me. The roller stand is essentially a sawhorse with a row of garage-door rollers mounted 3 in. on center across the length of the crossbar. I made my stand out of 2x4 stock, screwed and glued together for rigidity and braced at the top with plywood gussets. With one of these stands beside my table saw, I can push a sheet of plywood through a cut with little resistance. ***—Lewis A. Locke, Ellenburg, Wash.***

Chapter 4: Saws

Rip fence

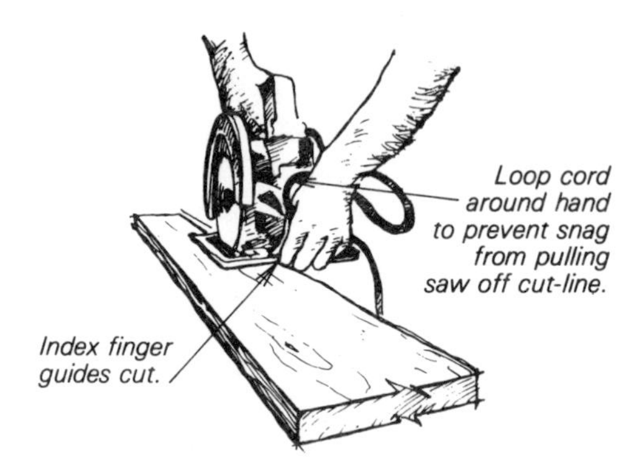

Most portable circular saws come with a rip fence that gives an accurate cut, but having one attached to your saw can be very inconvenient. Usually, you are switching back and forth between crosscutting and ripping, so you end up ripping freehand. The problem is how to make a true, straight rip.

As long as you have a saw with a foot that extends completely around the blade, you can use the fingers on your left hand (if you are sawing with your right) as an accurate rip guide. Hold the foot plate firmly at the front, allowing your fingers to extend below the plate enough to provide a pressure guide against the wood.

If your reaction is "that's dangerous," study the situation with the saw unplugged. You have a solid two-handed grip on the saw, and your fingers are well clear of the blade. If you also loop the cord around the thumb of the hand that's acting as the fence, you'll know if the cord catches on something before the saw is pulled off the cut.

—Kendall Gifford, Putney, Vt.

Stair-button rip fence

When I need an accurate rip on a job site without a table saw, I use my stair-gauge fixtures. I clamp them on the front and back of my circular-saw baseplate, equidistant from the blade. This requires a saw with a flat baseplate. Measure the distance from the blade to the stair buttons just as you would with a rip fence, and then make the cut normally. With this quick setup, I get table-saw accuracy.

—Gred Gross, Wooster, Ohio

Easy-glide saw table

For one reason or another, we all end up having to trim doors with a circular saw. Despite using precautions such as guides and tape to protect the work, many saws leave an unsightly streaking along their path, spoiling an otherwise unblemished job. Well, my saw was one of the guilty ones.

To avoid this problem, I now place a continuous layer of clear contact paper on the base of my circular saw. I cut out a slot for the blade and leave a ⅛-in. overhang at the rear to act as a pull tab. The contact paper has a clean, smooth surface that glides easily over a workpiece without leaving a trace. When the contact paper gets worn and needs replacing, it comes off easily with a tug on the pull tab.

—MacGill Adams, Anchorage, Alaska

Durable sawblade bushings

Several saws with ½-in. arbors use blades with ⅝-in. dia. centers—a plastic bushing makes up the difference. These lightweight bushings can be damaged pretty easily, especially when they are mounted in a circular saw. When our last one died, we hunted around for a more durable bushing and found it right on the job site—a ring of type L, ½-in. copper water-supply tubing.

Not only does a copper bushing last longer, but you also have some control over its length. A slightly longer bushing (about twice the blade thickness) makes mounting and positioning quicker. Don't let the elongated bushing contact both saw-blade compression washers, or power transmission to the blade will be reduced.

—Samuel M. Hoagland, Farmington, Maine

Miter-box medicine

One of the few drawbacks of the Makita power miter saw is that the space between the platen and the base gets clogged with sawdust, making it difficult to change the angle of the saw. To help solve this problem, cut a wood insert for the saw slot. Shape a small piece of pine to fit the radius on the bottom of the slot, and press-fit it in place. Then use the saw to cut a kerf in the insert. This trick will keep the sawdust out of the space under the platen, and as an added benefit, it also will reduce tearout on the bottom of your cuts.

—Ken Kellman, Ben Lomond Calif.

Cutting vinyl and aluminum

I've found a much better cutting tool than my circular saw for the occasions when I have to cut vinyl or sheet metal—a 100-millimeter disc grinder. It's a lot more maneuverable than a saw, and the blade and cutline are easier to see.

For sheet vinyl, I use a masonry blade in the grinder. It actually burns away the material and produces a smooth cut every time, regardless of the outdoor temperature.

For cutting aluminum, steel roofing or flashing, I replace the masonry blade with a metal grinding disc. And this is one job when I really do wear my goggles.

—Jim Billman, Sigourney, Iowa

A chopsaw table

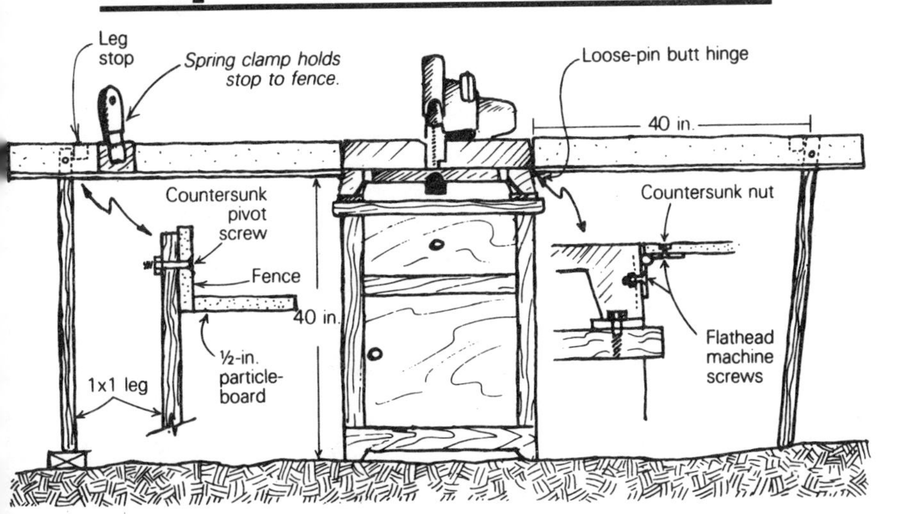

An article on power miter saws made me think about my own setup for the tool. As shown in the drawing above, I mount my 10-in. Makita on a cabinet that has a drawer and some shelves. This storage space is for stashing blades, sandpaper and other paraphernalia.

The heart of the system is a pair of extension wings that are attached to the saw with three ½-in. loose-pin butt hinges. The wings are built of ½-in. particleboard. They are about 5 in. wide and 48 in. long, with a 4-in. high fence that is edge-butted, glued and screwed. The hinges are attached to the saw's cast-iron table and the wings with flathead machine screws. The other ends of the wings are each supported by a 1x1 leg that is attached to the wing fence with a single countersunk machine screw. A glued-on stop keeps each leg from opening too far.

When I arrive at a job site, I first set up the saw and its cabinet, and then I slide the hinge pins in place to connect the wings. To get them level, I eyeball down the wings and across the saw table. I adjust them either by shimming the legs or swinging one of them back out of plumb.

This arrangement is very simple to adjust, and it's also portable. The 10-ft. long work surface gives me enough room to handle just about any board. Its 40-in. height is easy on my back, and the wing fences are good places to clamp stops for multiple cuts under 60 in.

—Bruce Donegan, Ridgway, Colo.

Born-again blades

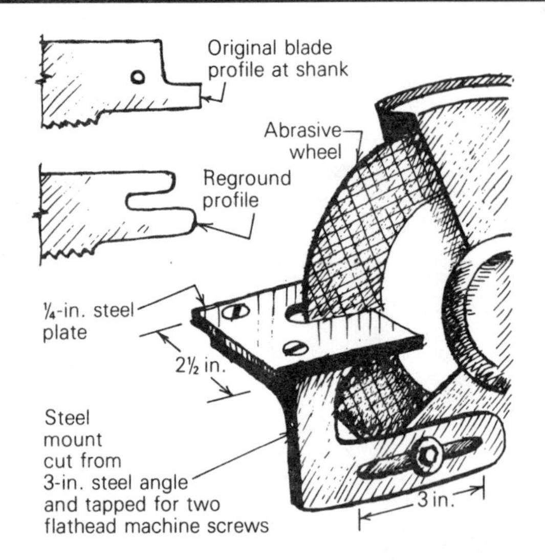

One of the most essential remodeling tools is the reciprocating saw, but remodeling work is very hard on their blades. Here are two tricks I use to prolong blade life.

First, I use 12-in. flexible blades whenever possible and do the cutting with the teeth closest to the tip. When the teeth wear down, I simply clip off the worn portion of the blade with tin snips and go back to work.

Second, and more important, I reshape broken shanks using a metal-cutting abrasive disc mounted on my bench grinder as shown in drawing above. Outline the shape of the original blade on the broken shank and grind away the excess metal with the abrasive disc until the new shank fits snugly inside the saw's shoe. Then cut a slot in the blade just wide enough to accommodate the pin and retaining screw.

I've found innumerable uses for the bench grinder with this wheel installed. It will accurately cut a variety of materials from spring-steel roll pins to rubber hose. A word of caution: Check the r.p.m. rating for the blade and the grinder to make sure they are compatible.

—Philip Zimmerman, Berkeley, Calif.

Resawing on site

If you've ever used resawn wood as a finish material, you know how troublesome it can be to run short by a few boards. The bandsaw operator at the lumberyard is never very happy to get an order for two or three pieces of custom-cut stock, and it can be a real hassle to coax him into putting aside larger jobs to fill your request.

The last time this happened to our crew, we solved the problem with a reciprocating saw running a dull, multi-purpose blade. We secured the work to a pair of sawhorses, and ran the blade flat against the finished side of the wood. The surface texture can be varied by changing the angle of the saw, the number of teeth on the blade and the speed of its reciprocating action. Wide boards can be textured with a longer blade worked from both sides.

—Ernie Alé, Santa Ana, Calif.

Quick square cuts

A technique I find useful for making an accurate crosscut with a circular saw is to place a framing square on the work and scribe a pencil line at the desired dimension. Then, because I'm right-handed, I slide the square to the left, away from the scribed line. Next I place the saw shoe against the tongue edge of the square and move the two as a unit until the blade lines up with the cutline. With a firm grip on the square, I start up the saw and use the framing-square tongue as a fence, as shown above. I get accurate, table-saw quality cuts this way every time.

—Randy Cofer, Boyds, Wash.

A carbide tip

I go through a lot of steel circular-saw blades remodeling and building additions. I hate to pay $2 a blade for sharpening, and then wait a week to get the blade back from the shop. But doing the sharpening myself takes more of my time than I'd like.

The answer for me is to use carbide-tipped blades and sharpen them myself with a diamond whetstone. Unlike the teeth on a steel blade, the carbide teeth on my blades are a simple series of flat planes that are easily reached with a small, flat stone.

The stone I use is a plastic block 4¼ in. by 13⁄16 in. by 7⁄32 in. Diamond crystals are bonded to a nickel matrix that's embedded in the plastic. Mine was manufactured by Diamond Machining Technology, Inc. (34 Tower St., Hudson, Mass. 01749) and cost $16.50 when ordered from a tool catalog. These diamond whetstones come in several sizes and are also just the ticket for those impossible-to-sharpen stainless-steel kitchen knives.

Although I take my blades to be professionally sharpened every third time, the touch-ups have paid for the stone many times over. ***—Angelo Margolis, Sebastopol, Calif.***

Warning zone

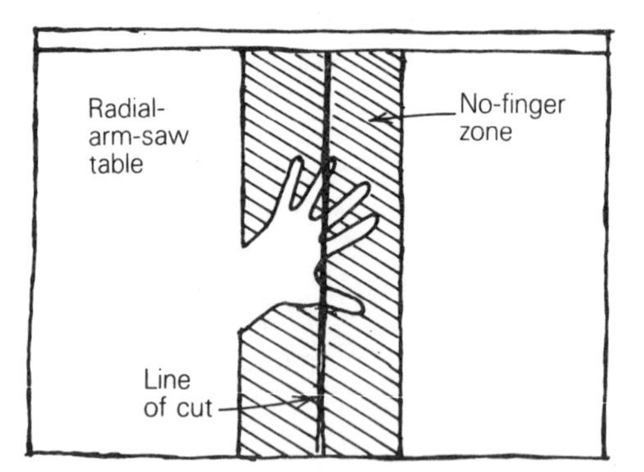

After my last run-in with my radial-arm saw (finger remains, but with ugly scar), I decided that I needed a graphic reminder to use special care when my fingers get close to the blade. With the saw unplugged and to the side, I placed my hand down on the saw table, right in the middle of the cutline. Then I outlined my hand with a red marker and drew diagonal lines to represent the danger area. So far this no-finger zone has been very effective at reminding my brain to pay close attention to the placement of my hands while I'm using this tool. *—Jim Fish, Dale, Tex.*

Cutting through roots

Digging trenches for pipes or foundations can be a miserable job if you're working around trees. I recently had to install a large drainage system that passed through the roots of two massive pines, and I finally discovered a tool that really works on the roots—a Sawzall with a 12-in. pruning blade. Throw down the axes, hatchets, picks and handsaws, and slice through the roots with this versatile tool. In damp soil, protect yourself against shocks with a GFCI-equipped extension cord. *—David Bainbridge, Berkeley, Calif.*

Custom jigsaw base

I got tired of scratching my woodwork with the base of my Bosch jigsaw so I replaced it with a wooden one. I used a piece of ⅜-in. thick maple, 3 in. wide and 5 in. long. The larger base gives more stability, and the wood damps vibration. *—Will Milne, San Francisco, Calif.*

Non-clogging sawblades

Have you ever tried cutting through a composition-shingle roof, only to find that your reciprocating sawblade has become totally fouled with tar and refuses to cut? It's frustrating to stop every 6 in. or so to unclog the teeth. I solved the problem by switching to a plaster-cutting blade. Its symmetrical teeth are non-clogging and they cut in either direction. Since they catch more often on the forward stroke, invest in the sturdiest bimetal blades. In my experience, Lenox blades (American Saw Mfg. Co., Box 504, East Longmeadow, Mass. 01028) have proven to be the best—they don't snap off, and can be straightened out if they bend. *—William H. Brennen, Denver, Colo.*

Built-up roof blade

I've installed a lot of skylights in tar and gravel roofs. As a consequence, I've gummed up plenty of circular-saw blades trying to burn through the built-up roofing. Since this kind of cutting ruins the blades anyway, I decided to modify a blade so that it would work better.

I selected my dullest combination blade, and set the teeth way out—¼ in. or more. This radical set allowed the blade to clear itself with each revolution, and worked even better than I'd hoped. —***Bob Whiteley, San Rafael, Calif.***

Close shaves on the table saw

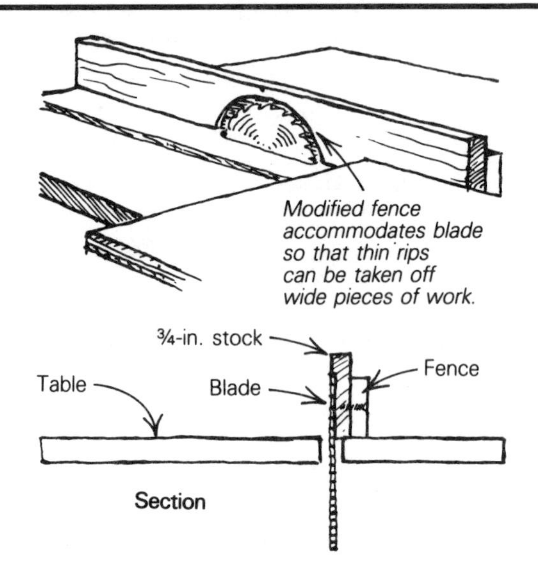

Whenever I do finish carpentry on site, I use a small, portable table saw for everything from trim to cabinet work. Unfortunately, the diminutive table limits the width of the work that can be passed by the blade. Sometimes I have to take minuscule rips off the edge of a large piece of plywood, so I devised the modified fence shown in the drawing above to allow close shaves on the little saw.

To modify the fence, I screwed a ¾-in. thick, knot-free piece of wood to the blade side of the fence, making sure that the screws were well away from the path of the sawblade. Then with the blade lowered below the level of the table, I positioned the wood so that it was flush with the outside edge of the blade. I turned the saw on, and slowly raised the blade to full height, cutting a half-moon void in the wood fence.

I use this fence along with an 80-tooth carbide blade to take 64ths off wide stock.

—Jeffrey S. Janssen, Oakland, Calif.

Fixed protractor

I use a saw protractor for many of my cut-off needs. It saves the step of drawing the cutting line on the material, and is highly accurate. I find it particularly useful for cutting bevels such as scarf-jointing 2x fascia boards.

Because most of my cuts are square, I have altered my protractor for long periods of use at 90°. To do this I set the protractor with a reliable framing square, tighten the wing-nut clamp, and drill a pilot hole through the calibrated arc into the arm. A self-tapping sheet-metal screw holds it at a perfect 90°, ready for all square cut-off needs. The screw is easily removable for angle cutting.

—Ralph Dornick, Altoona, Pa.

Cleaning sawblades

Pizza pans, which come in various diameters with a standard depth, are ideal for soaking circular-saw blades in chemical remover to get rid of pitch. Very little fluid is needed, and after an overnight soak the residual fluid can be poured back into its can to be used again. The blades come clean with little effort. ***—Mike Denman, San Jose, Calif.***

Chapter 5: Tool Tips

Dustless drilling

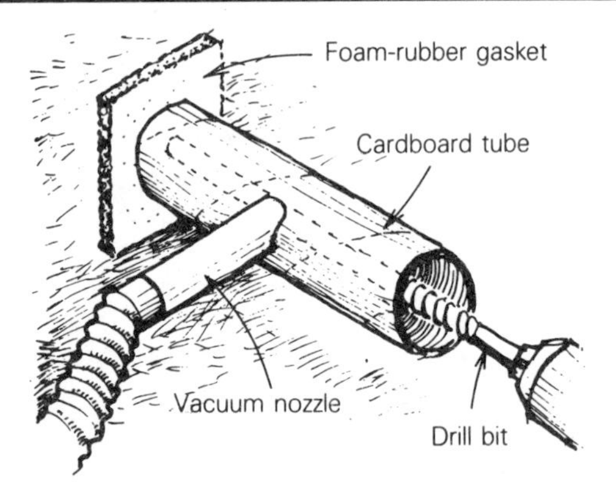

After our company had painted and decorated the interior of an ornate 19th-century church, we had to install a new sound system. This meant drilling through a 20-in. thick plaster and brick wall located directly over a gilded ornamental plaster arch. To prevent a disastrous mess of fine brick powder filtering down upon the arch, I devised the vacuum fitting shown in the drawing to contain and collect the dust.

The fitting consisted of a 3-in. diameter cardboard

mailing tube about 8 in. long, and a thin gasket of foam rubber to seal the space between the wall and the end of the tube. I cut a hole into the side of the tube to receive a vacuum hose, which I taped to the fitting for an airtight seal. As we drilled into the wall with a heavy-duty hammer drill and a 30-in. masonry bit, the dust was contained within the cardboard tube and immediately sucked up by the vacuum. The hole was drilled without a trace of dust reaching the arch.

—Jeff Kerbeykian, Rego Park, N. Y.

Stringing wedges

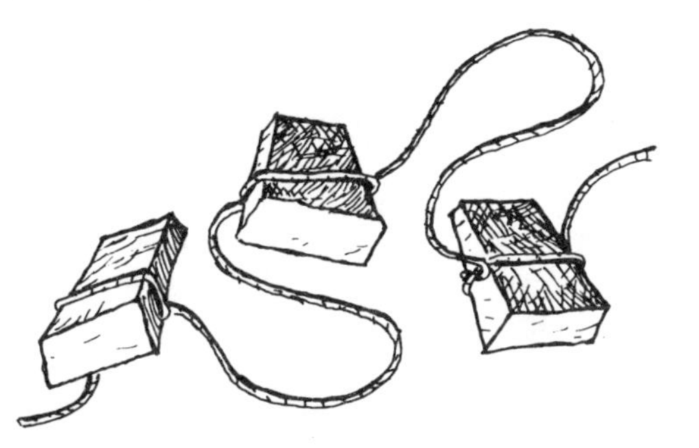

I use steel wedges to split sandstone boulders into building blocks. But our leafy groundcover sometimes makes it tough to find the wedges after a cut. To make sure I don't have to spend time on my hands and knees hunting for them, I tied my wedges into a necklace like the one in the drawing.

—Gary Loomis, Mendocino, Calif.

Magnet helper

A magnet can be a very handy tool around the shop or job site. When I'm working with brads or tacks, I use a magnet to keep them all concentrated in one place. The little nails will stick out like tiny thorns, making them easy to grab between two fingers.

Whenever I have to disassemble a tool and I'm liable to misplace a part in the surrounding sawdust, I get out a magnet to manage all of the parts. Not having to look for lost screws saves a lot of time.

—Ron Davis, Novato, Calif.

Mixing cap

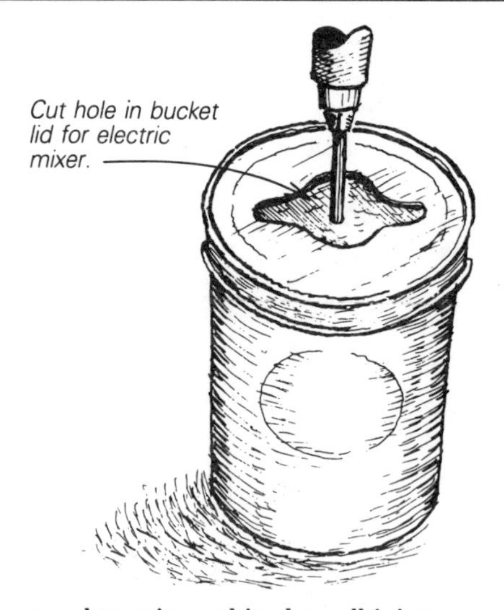

When you need to mix or thin drywall joint compound with an electric-drill mixer, save yourself some cleanup time by using a mixing cap on your 5-gal. bucket. Make the cap by cutting a hole in the lid of an extra joint-compound bucket, as shown in the drawing above. Make the opening in the center of the lid large enough to add water or joint-compound powder as you mix, and you're in business.

—William Barstow, Davis, Calif.

Epoxy stirrer

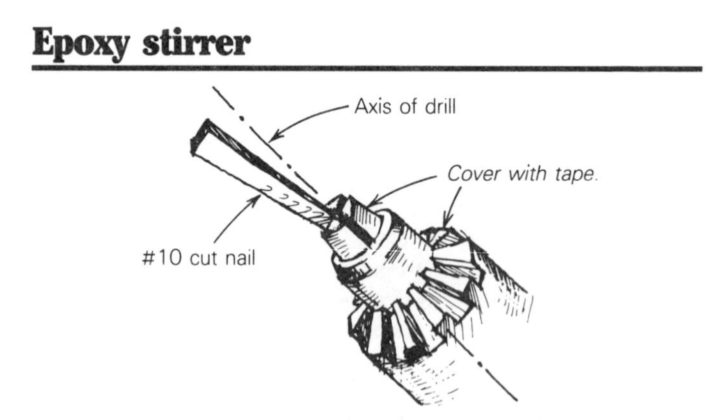

I frequently use liquid epoxies to repair and restore old windows and doors in my house. Since thorough mixing of the epoxy is essential, I have tried a variety of jury-rigged stirrers with my variable-speed electric drill.

I've gotten my best results by using a #10 masonry (cut) nail, chucked between two of the three jaws of the chuck, as shown in the drawing. Gentle stirring of smaller volumes is best achieved by keeping the nail just a bit off axis. I get a more vigorous stirring action by moving the nail farther off its axis. In order to protect the drill chuck from epoxy spatter (which will quickly freeze the chuck), I wrap the end of the chuck with tape. A soda can, cut in half, is a very convenient container for the volume that I usually mix, and it works well with the cut nail.

—Jeff Kerbeykian, Kew Gardens, N. Y.

Freeze alarm

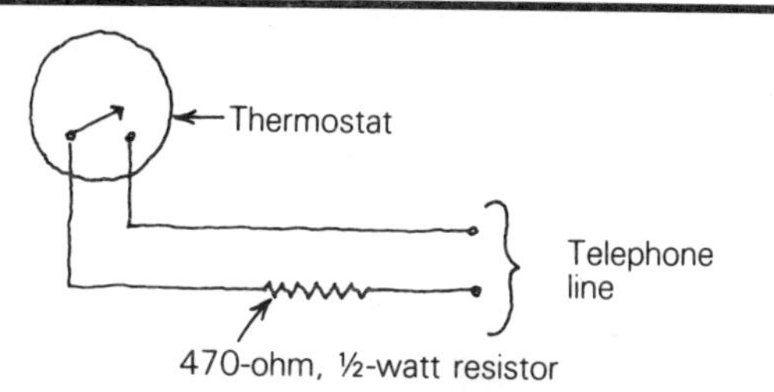

It's often necessary, or at least convenient, to leave the heat running in a house that's being remodeled or in the latter stages of construction. A warm house means you don't have to drain water pipes and radiators during the winter, and footings are protected from frost heaves. Because I'm primarily a weekend remodeler (and a worrier as well), I've often made needless trips to a site during the week just to assure myself that the furnace hasn't quit. No heat for several days at -20° is the stuff of nightmares and insurance hassles.

I got tired of all the unnecessary driving, so now I call a project to check on the heater. I installed a standard heating thermostat across the telephone line with a 470-ohm resistor (Radio Shack part 271-019) in series, as shown in the diagram above. I set the thermostat at 40°F, or whatever minimum temperature is permissible. If the house temperature falls below it, the thermostat contacts close, placing the resistor across the line. This condition simulates an off-hook telephone and causes a busy signal if the number is dialed from any other phone. To verify house temperature, I dial the number. If it rings, everything is okay. Also, when the house is warm and the contacts are open, the site telephone sharing the same line functions normally.

I use a thermostat with mercury contacts, which requires a firm, level mounting position to be reliable. Older thermostats with mechanical contacts are not so sensitive to position, and would probably be better for this application. Their minimum temperature, however, is often 50°F or 55°F.

—Karl Juul, Slingerlands, N. Y.

Bearing-over bit

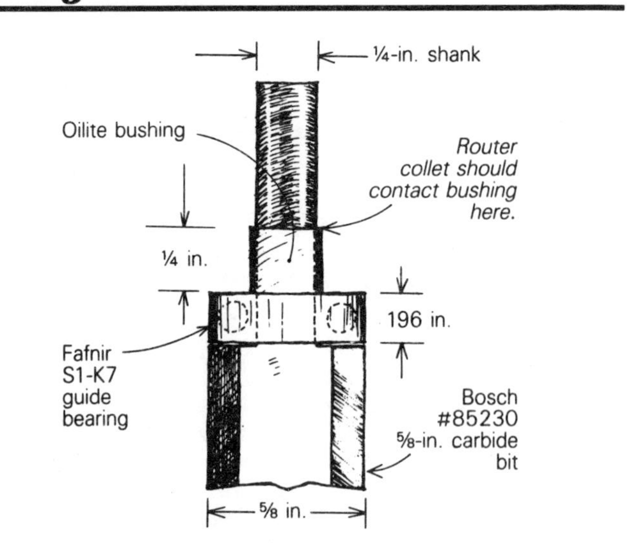

I recently found myself needing a two-flute router bit with a bearing mounted above the cutters. My supplier didn't have one, and I didn't feel like waiting for the postman. So I pieced one together using readily available parts.

I started with a two-flute carbide bit (a Bosch #85230, ⅝ in. in diameter). Next I located a guide bearing of the same shaft size and outside diameter as the cutters (Fafnir S1-K7), and a short Oilite bushing (¼ in. i.d. by 5⁄16 in. o.d. by ¼ in. high). The bearing slips over the shaft, and the bushing keeps the bearing close to the cutter – provided that the bushing is touching the collet. This contact is important because it eliminates any up-and-down play in the bearing.

The system works quite well. The parts cost about $22, and they were all easy to find.

—Josh Rothman, Toledo, Ohio

Push-broom upgrade

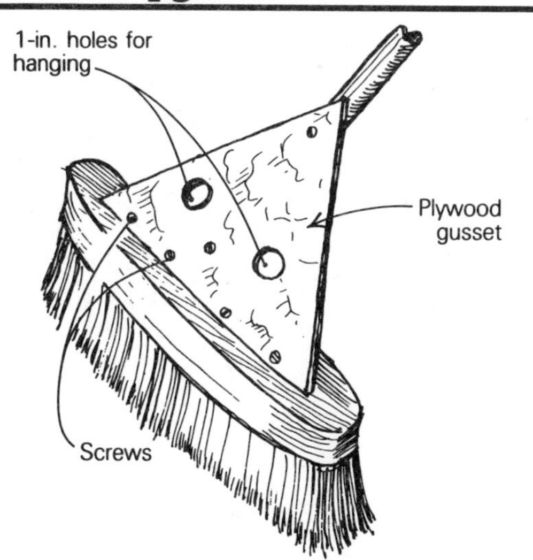

I don't even wait for my push broom to break before I strengthen it. When it's brand new, I screw the base of a triangle of ¼-in. or ⅜-in. plywood to the top of the broom head, and the top of the triangle to the handle. This gusset keeps the broom together longer, and the added weight helps it sweep better. ***—Michael Sweem, Downey, Calif.***

Head lights

I do quite a bit of remodeling and termite work, and I've found that my most valuable tool is my headlamp, purchased at a local sporting goods store. It's simply an encased bulb with a reflector fixed to an elastic strap that fits around the head. A separate battery pack attaches to my belt, and a small cord connects the two. It uses four penlight batteries and is very compact.

Before I thought of using this amazing gadget, I had to devote one of my hands to carrying a flashlight whenever I went under a house to do inspections. If tools had to be carried, I was limited to half my capacity.

No more training clamp lights or drop lights on job areas; no more shadows to fight with. All I have to do now is look at what I want to work on and it's illuminated.

I got my headlamp from the REI Co-op. For their catalog, write Box C-88125, Seattle, Wash. 98140.

—Alan Forbes, Oakland, Calif.

Wooden wheelbarrow

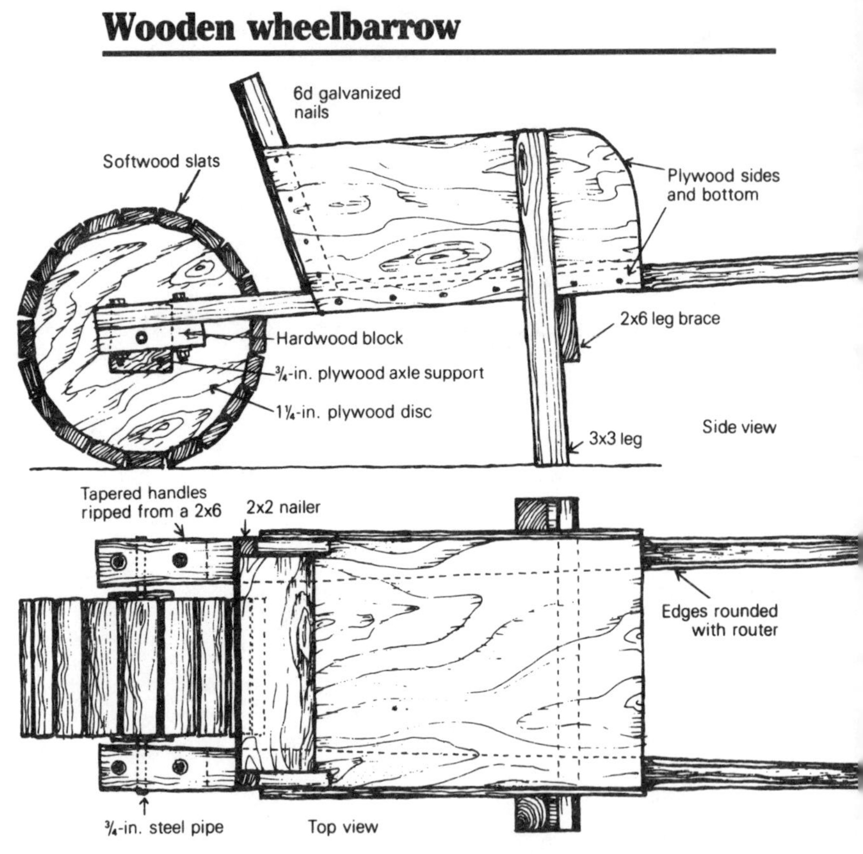

A couple of years ago I watched a neighbor struggle through the mud with a load of firewood in his wheelbarrow. The wheel was too small. Movement was all but impossible because there simply wasn't enough surface area to keep the front end up. Not long afterward, I designed and built the wheelbarrow shown in the drawing above. Handles for the wheelbarrow were cut from a 6-ft. length of full-dimension 2x6, ripped diagonally so that the portion over the wheels (the area that usually breaks first) was about 4 in. wide. The handle area was rounded a bit with a router. The wheel was fashioned from two 20-in. discs of 1¼-in. plywood. These were covered with softwood slats about 1¼ in. thick and 12 in. long.

The axle was formed from a section of ¾-in. steel pipe about 22 in. long. This was inserted through the center of the wheel and fastened in place with epoxy and a couple of scraps of ¾-in. plywood on each side. Bearings for the unit were made from hardwood blocks about 2 in. by 4 in. by 10 in. These were drilled about 1⁄33 in. oversize and bolted in place to take the axle. I use a few drops of 90-wt. gear oil on the axle from time to time as a lubricant.

The rest of the construction is fairly straightforward. I used a few scraps of ½-in. by ⅝-in. plywood for the sides and bottom, fastened with glue and 6d galvanized nails. I used 2x2 blocks in the corners for nailers. A couple of coats of Cuprinol gives protection from the weather.

The completed unit cost me less than $4 and took three hours to build. It is light, tough, holds a lot, and rolls more easily than commercial units. Its large wheel will carry tools and building materials through mud and over soft lawns without sinking in and leaving marks.

Other than low cost and simple construction, I think dependability is this wheelbarrow's biggest virtue. It will never have a flat tire.

—Mark White, Kodiak, Alaska

Drill-bit depth gauge

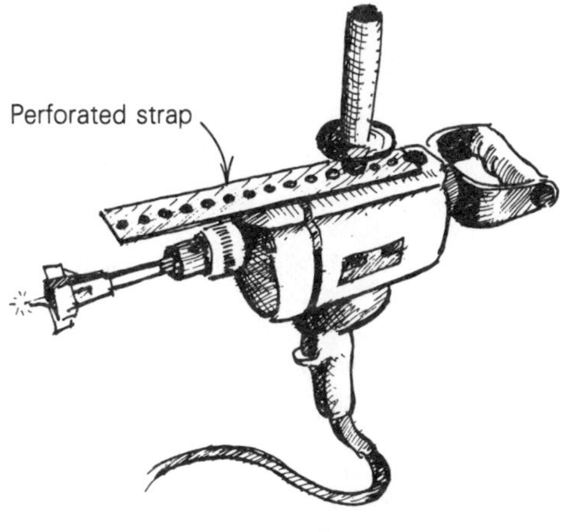

I'm a timber framer, and often need to drill large-diameter holes of various depths in big beams. I keep several lengths of perforated steel strap (purchased at a local hardware store) to use as depth gauges on my drill. As the drawing shows, the auxiliary handle holds the strap in place. The numerous, regularly spaced perforations allow a wide range of adjustability. One caution: Put a longer bolt in the auxiliary handle—I stripped the threads in my drill because not enough of the bolt was engaged during operation.

—Jack Winegar, Worth, Ill.

Boom-box deflector shield

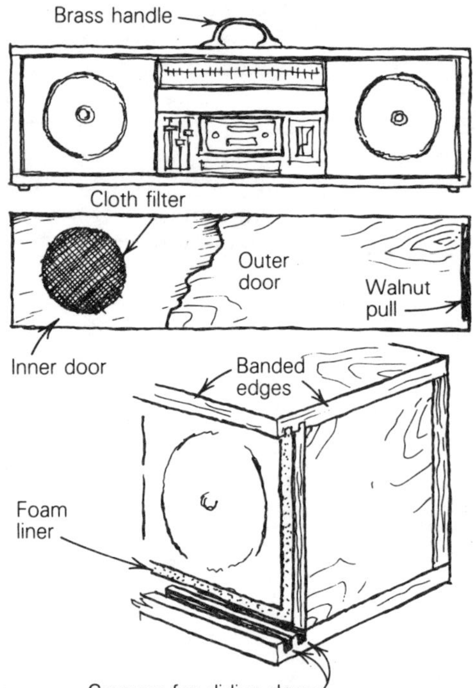

On my job site, the work radio pulls more than its share of the load. It's the first thing plugged in and the last thing unplugged. But job sites are dangerous places for radios. They are dropped, hit by falling tools and debris and slowly suffocated by dust. To protect my new radio from these evils, I housed it in a box made of ½-in. maple plywood, shown in the drawing above.

The front of the box has a pair of grooves that house ⅛-in. maple plywood doors. The outer door is solid, while the inner door has cutouts that correspond with the speakers. I covered the cutouts with a tightly woven cloth to keep dust out of the speakers. Both doors have walnut pulls at one end. I lined the inside of the box with foam to keep the radio from rattling around, and I put four maple feet on the bottom of the box to minimize scratching. Access to the power cord is through a hole in the back of the box. Besides protecting my $200 radio, the box has impressed my clients with the thoroughness of my work.

—MacGill Adams, Anchorage, Alaska

Dust-mask filters

Disposable paper dust masks can be used to filter thinners and light finishing oils. To support the mask as the liquid is poured through it, I put the mask in a funnel. Now that I know what to do with the dust masks besides throwing them away, I change my masks more often and I feel less wasteful about it. The masks work better than paint filters, especially on expensive solvents like lacquer thinner.

—*Michael R. Sweem, Downey, Calif.*

Little bits

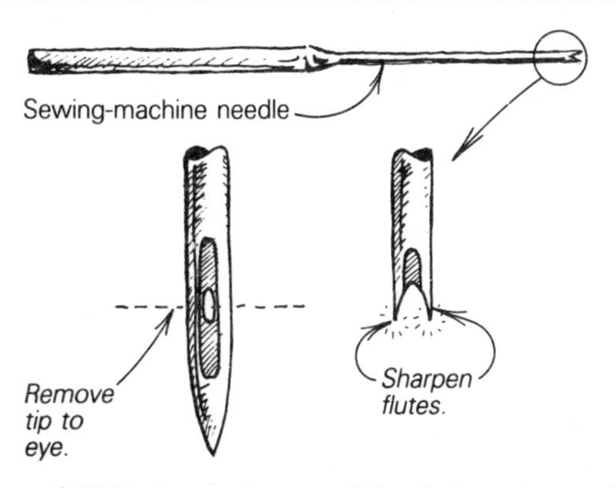

To make a drill bit for setting small brads in oak or other hardwoods, grind the tip of a sewing-machine needle to the middle of the eye, as shown in the drawing, and sharpen the resulting flutes as you would a twist drill. A pin vise makes it easy to hold the needle as you grind it. Even broken needles can be salvaged, and they bore excellent pilot holes—even in thin metal.

—*Jeffrey D. Taylor, Newport, Ore.*

Replacement handles

I have broken innumerable handles on hammers, mauls, axes and sledges, and somehow the replacement handle never fits quite as well as the original. It usually comes loose after a period of use, no matter how hard I drive in the end wedges. My solution to this problem is to use five-minute epoxy to glue the handle to the head. I clean out the hole in the hammer head, fit the new handle to it using a rasp, and then liberally coat both the handle and the hole with the two-part epoxy glue. Then I drive the handle into the hole, set the wooden and steel wedges, and allow the glue 24 hours to cure. The epoxy fills any gaps between the handle and the steel. I've never had a replacement handle come loose using this technique. ***—Craig Stead, Putney, Vt.***

Overhead outlet

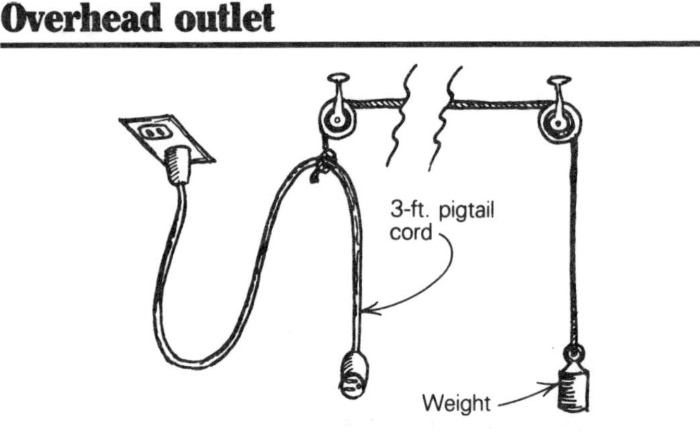

I installed an outlet directly above my workbench, and I've got a 3-ft., 12-ga. pigtail cord plugged into it. Near the female end I tied a clothesline cord that runs through two pulleys. The other end of the clothesline is tied to a weight (drawing, above), which keeps the pigtail just within my reach—about 7 ft. above the floor—and gives me a couple of extra feet of extension cord to use when I need it. The best part of this whole arrangement is that I can work all the way around my bench without having to untangle myself from the power cords when I'm using tools like drills and routers. I also have the overhead outlet on a switch, which lets me turn off the juice easily whenever I need to change bits or blades.

—Dale Hopwood, Bakersfield, Calif.

Hybrid handles

In carpentry circles the debate still rages as to which framing hammer is the best. Purists defend the wooden handle for its superior ability to absorb vibration, as well as the fact that most of the weight is located right where it's needed—at the head of the hammer. With steel handles, the weight is distributed along the length of the hammer. But wooden handles need frequent replacement, while some of the steel-handled brands are guaranteed for life.

For the real wildmen on a crew, only the framing hatchet will do. Every carpenter I know who can sink a 3½-in. spike with one hit uses a framing hatchet. This is due in part to the hatchet's superior handle design—it's neither round nor oval, and its shape matches the grip of the human hand. In addition the handle broadens out at the end, which encourages a carpenter to use all of the leverage that an 18-in. handle gives instead of choking up. Unfortunately, using a framing hatchet usually necessitates carrying around a nail-puller.

Perhaps the best of both worlds is my hybrid, which combines a framing-hatchet handle with a 28-oz. hammer head. The hatchet handle is shaped differently from the standard handle, so I had to whittle down the end before I wedged on the head. There were still a few gaps between the handle and the head, but I used these to advantage by filling them with fiberglass resin. Then I saturated some fiberglass mesh with more resin, and wrapped it around the juncture of the head and the handle. Now my hybrid hammer is one of the strongest, most durable framing hammers I've seen. ***—Art McAfee, Edmonton, Alta.***

Starting a siphon

Every once in a while somebody has to use a large-diameter siphon to bail water out of a hole, to drain a tank or to empty a boat full of water. In my case, I had a 14-ft.-deep hole that I needed to drain so that I could install a wooden enclosure around what would eventually be a surface well. The flow into the hole was about 40 gallons a minute—substantially more than a garden hose could handle.

I had on hand a section of 2-in. flexible plastic pipe, about 70 ft. long. I drilled a small hole near one end of the pipe for a piece of wire. Then I tied a chunk of heavy iron to the wire so that one end of the pipe would quickly sink to the bottom of my hole when it came time to start the siphon. I laid out the rest of the pipe down the hill from the hole.

Now the task at hand was to start the water flowing. Anyone who has ever attempted to start a siphon (especially a large-diameter one) knows that it can be an exasperating experience. To get this siphon going, my son and I ran a long wire through the length of the pipe. Then we tied a small-diameter rope to the wire, and pulled it through the pipe. To the rope dangling from the end to be submerged, I knotted a wad of rag from an old shirt. When I pulled on the rope at the opposite end, the rag entered the pipe, making a seal. I then threw the weighted end of the pipe into the water, and I held onto the pipe while my son made a mad dash downhill with the rope, pulling the cloth plug and a column of water behind it through the siphon. The water started flowing immediately, and the hole was drained in about two hours.

—Mark White, Kodiak, Alaska

Extension-cord garage

The most convenient way that I've found to deal with unruly extension cords, especially if you move your work stations around, is to carry them in an old joint-compound bucket. As shown in the drawing above, I cut a hole near the bottom of the bucket to accommodate the prong end of the cord. In addition to holding a long cord, there's plenty of room in there to carry a couple of power tools as well.

—Bryan Humphrey, Denver, Colo.

Screw-eye driver

The next time you've got to install cup hooks or screw eyes, get out your old brace and bit and make the job easy. A brace with a two-jawed chuck will get a good bite on the screw eyes, and a starter hole in harder woods will help get things going.

—Mark Quigley, Boston, Mass.

Blueprint basket

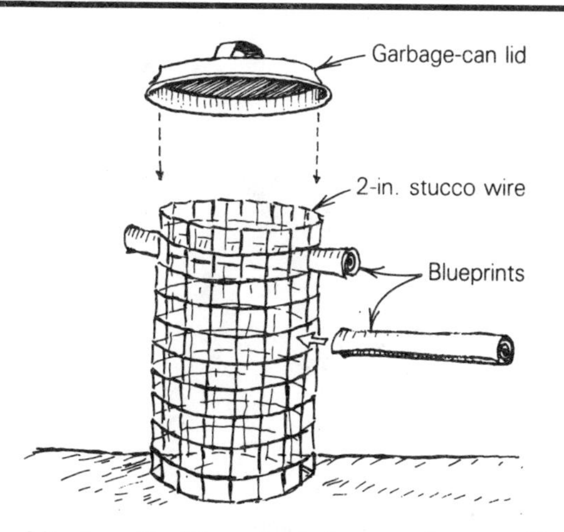

I'm an architect and builder who is just starting an office, and my equipment budget is slim. To keep the rolls of blueprints off my desk, I devised the basket shown in the drawing. It's made of 2-in. stucco wire that's been fastened into a cylinder. The diameter matches that of a surplus garbage-can lid. *—Jim Ramsay, Kelowna, B. C.*

Color-coded tools

Sometimes I've got so many power tools plugged into various outlets on the job that it becomes impossible to distinguish which cord goes to which tool. To make sense of this tangle of spaghetti, I bought five different colors of tape at my electrical-supply house. Then I color-coded my tools. I marked each cord with a wrap of tape near the plug, another near the midpoint of the cord and a third next to the tool. When I got more than five tools, I started putting two colors of tape on the cords.

This simple identification strategy saves me time when I need to unplug one tool and plug in another, or find a specific cord when they become entangled.

—Bill Haynes, San Francisco, Calif.

Slater's hammer

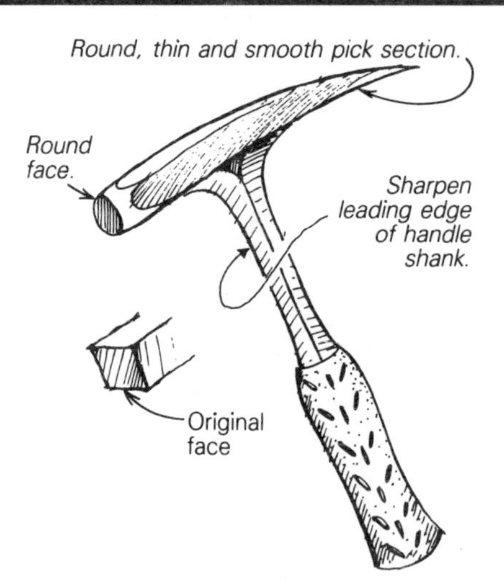

People who have worked with roofing slate know the value of a good slater's hammer for punching, breaking and trimming brittle slates. After using a traditional slater's hammer borrowed from the local tinsmith, I set out to find one of my own. I watched the classified ads, attended flea markets and tag sales, and shopped trade and builder's suppliers. Used hammers didn't appear, but I was pleased to find a high-quality, brand-new hammer with a leather grip. Pleased, that is, until I learned that they cost from $60 to $80.

An alternative tool was already in my kit, and it took me just 15 minutes at the grinder to modify it.

Estwing geologist's rock picks, with 14-oz. or 20-oz. heads and a choice of nylon or leather grips, cost less than $20. I modified mine by rounding the hammer face and pick sections to a more slender, smooth profile, as shown in the drawing above. It's especially important to remove the square corners on the hammerhead—they are liable to break off while hammering nails. I also sharpened the leading edge of the solid steel shank. The modified rock pick makes it a simple matter to punch a smooth hole in a slate, and the sharpened shank has trimmed hundreds of slates without need for further touch-up. I have never missed the small side claws on the traditional hammerhead, and I believe that the 14-oz. modified Estwing has a better feel than the originals. ***—Bob Sager, Bondville, Vt.***

Earplug storage

I use a chainsaw to make a living, building scribe-fit log houses in Southern Idaho. Some type of ear protection is a must in this business, and I use the soft foam earplugs that resemble long, miniature marshmallows. The trouble is, when I take them out and drop them into my tool pouch, they quickly become too grungy to re-use.

I solved this problem by putting a plastic 35mm film canister on my tool belt to act as a storage jar. I cut a slit in the pry-off lid, slipped a length of knotted leather thong through the cut, then tied off the thong on my tool belt. Now I've got a handy, clean place to keep my earplugs.

—Tom Balben, Teton Village, Wyo.

Modified spade bit

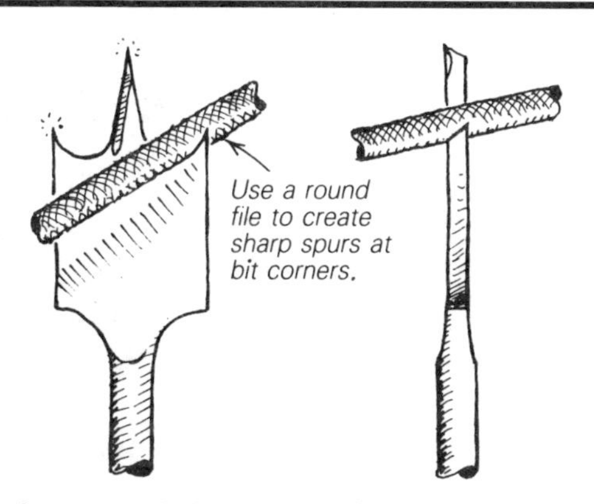

I commonly use spade bits to bore ½-in. to 1-in. holes in wood because they are cheap, readily available and easily sharpened. But in order to make a clean hole in wood, a spade bit has to be sharp. I keep mine tuned up by sharpening them with a round file—the kind used for sharpening chainsaw teeth. As shown in the drawing above, I file the inboard cutting edges of the bits to make them curved. This filing leaves a pair of sharp spurs that cleanly slice the wood fibers during a cut.

I have also used a spade bit modified in this manner to bore slightly oversize holes for dowels. I spread the spurs outward a little by using a ball-peen hammer and a vise for an anvil. After the holes have been drilled, I file the bit back to its original diameter. ***—Roger Niesen, Ellsworth, Maine***

Stone touch-up

I use a Carborundum stone for sharpening. But keeping it flat and getting it to hold its lubricant, despite repeated oil soakings, have been continual hassles. To solve the first problem, I grind the stone on a broom-finished concrete floor. This is fast, and accurate enough. To prevent the lubricant from soaking through, I immerse the stone in a bucket of water. When the stone is saturated, a light application of oil or kerosene will stay right on the top.

—Tom Law, Davidsonville, Md.

Eyedropper recycled

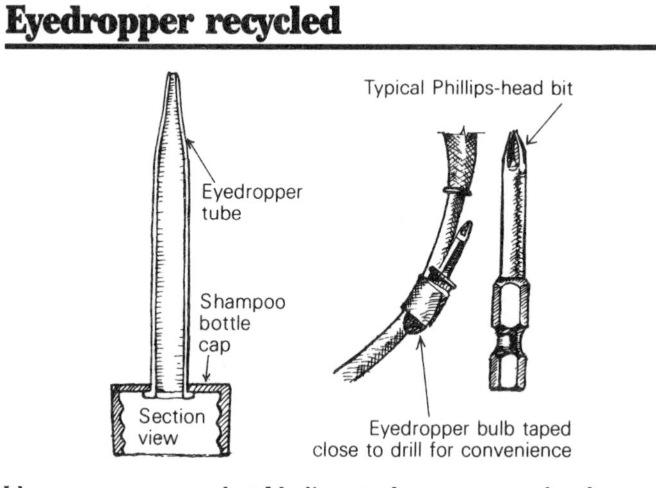

I have come upon what I believe to be a new use for the common eyedropper.

As a cabinetmaker/carpenter, I find myself often using my electric drill to drive screws. The Phillips-head bit I use was always being misplaced or falling to the bottom of my work pouch. I taped the bulb part of an eyedropper to my drill cord. It holds the bit quite well.

I use the tube portion to convert an old shampoo bottle into a fine glue bottle. I drilled a hole in the screw cap of the bottle and slid the tube into place; the hole was sized by testing the tube in a drill index and drilling the hole 1/64 in. smaller. This glue bottle can be any size and has a tip that will reach hard-to-get-at places.

—Mark Hallock, Arcata, Calif.

Increasing the power supply

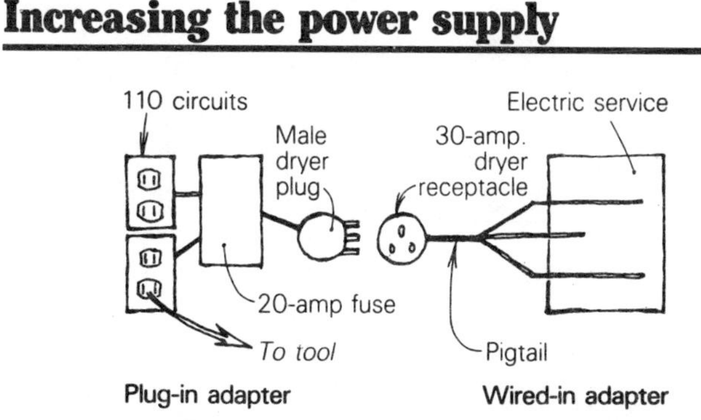

Antiquated wiring can be a real frustration during a remodeling job—under a load, compressors and table saws are always blowing fuses. When this happens, I do one of two things. If the house has an electric clothes-dryer, I plug an adapter I made up into the dryer's 30-amp circuit. The adapter is fused for 20 amps, and it gives us two separate 110-volt circuits.

If there isn't a dryer, I go to the electric service and tie a pigtail to one of the fused circuits. Our pigtail has a female dryer receptacle on its end that allows us to plug our adapter into it, as shown above. If you don't know your way around a fuse box, have the electrician on your job hook up the pigtail to the electrical service for you. Either way, it's a lot easier than running back and forth between your work and the service panel to change fuses or throw breakers for the course of a job. ***—Angelo Margolis, Sebastopol, Calif.***

Goggle defogger

My protective goggles would fog whenever I used them with a dust mask. My breath would sneak through the gap at the bridge of my nose, condense on the goggle lenses and make it impossible to see. Then I remembered a trick that I'd learned about shaving at the bathroom mirror. I rubbed the inside of the lenses with a little soap and water (mostly soap), and the problem cleared right up.

—Karl Riedel, Takoma Park, Md.

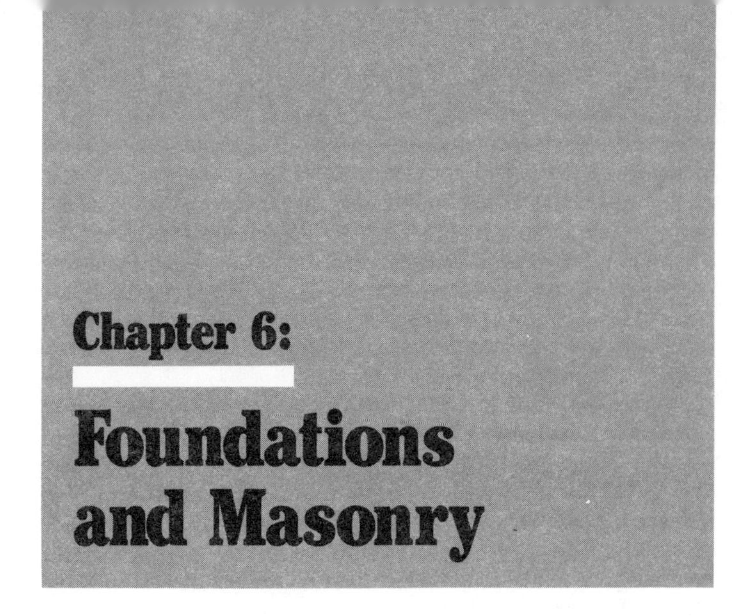

Chapter 6: Foundations and Masonry

Sliding dump-bed

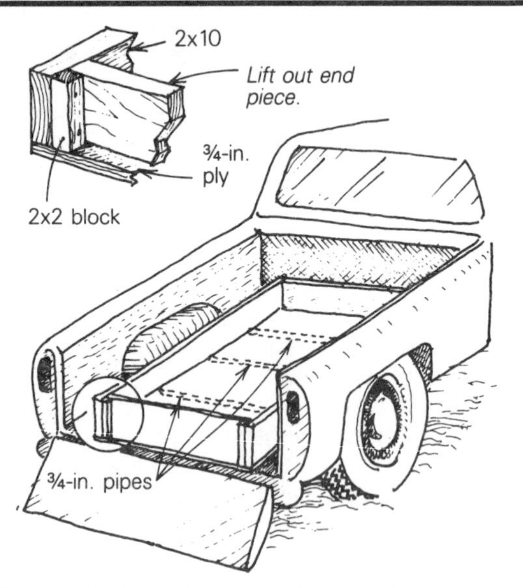

The mother of this invention is my hatred of shoveling aggregate out of my pickup truck. I built a shallow wooden box that keeps the stone in one place when I drive, and dumps it exactly where I want it when I get to the site.

The dump-bed box fits between the truck's wheel wells, where it rests on two or three lengths of ¾-in. pipe. When the dump-bed is empty, I block it up with a couple of 2x4s to keep it from rolling around while I'm driving. Before loading I pull out the blocks and make sure that the pipes are oriented as shown in the drawing. This dump-bed holds about a half-yard, and the loader operators I've worked with so far have been able to fill it without spilling a granule.

To unload, I drop the tailgate all the way down, back the truck up to the dump point and hit the brakes. The dump-bed rolls back, coming to rest with one end on the ground and the other on the back of the truck. I just pull out the end piece, and the rest of the aggregate can be pushed out of the dumper. ***—Al Dorsa, Christiansted, St. Croix, V. I.***

Pneumatic duplex nails

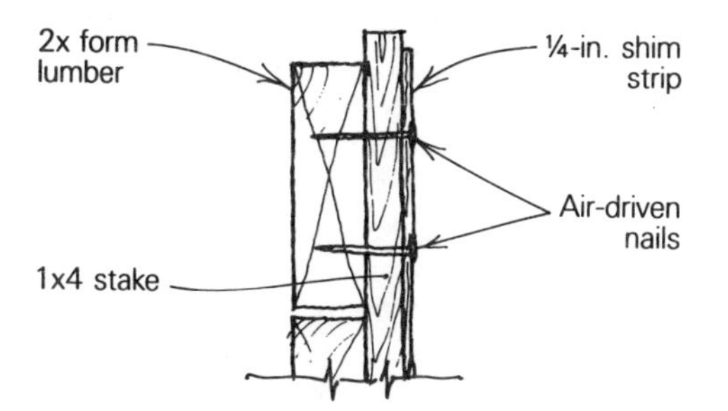

I don't know why somebody hasn't invented a nail gun that shoots duplex (scaffolding) nails. In foundation forming, pneumatic nailing increases not only efficiency, but accuracy as well. But trying to strip form boards where the nails have been driven home guarantees stakes broken off in the ground, concrete scarred by fighting the forms, and a lot of unnecessary work and frustration. Here's how I use a standard nail gun to get the advantages of both air-driven fasteners and duplex nails. Before I begin my foundation forms, I tack or staple ¼-in. shim strips to all my vertical stakes. Then I use my air gun to attach the stakes to the form boards. When it's time to remove the forms, I simply hack off the strips with the claw of my hammer and yank out the nails. ***—Michael Gornik, Nevada City, Calif.***

Tunneling under slabs

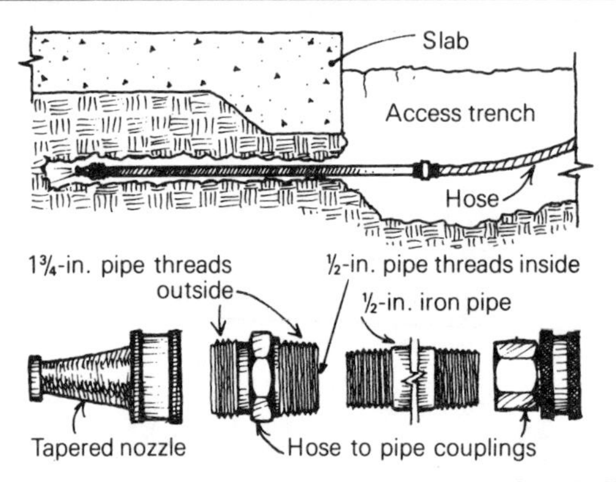

Have you ever needed to bury a water, gas or electric line and found your path blocked by a concrete slab? Here's a hydraulic method for making a small, accurate tunnel under such an obstacle, using a garden hose, iron pipe and about $5 worth of common brass fittings from your hardware store.

Excavate a trench to the necessary depth, on both sides of the slab, and assemble the pictured fittings. It is important to maintain a level course under the slab, so be sure your trench is long enough to allow the pipe to remain level while the tunnel is being cut. If access is limited, short sections of pipe may be added with couplings as the tunnel gets longer. Sometimes it is best to work from both sides and meet in the middle.

The tapered nozzle delivers water at a very high velocity and quickly erodes the soil in its path. Adjusting the flow of water will control the diameter of the tunnel.

*—**Bruce Goodell, Oakland, Calif.***

Flagstone template

To keep waste to a minimum when I lay a floor of irregular flagstone, I first lay the large pieces of stone without cutting any of them. Then I try to match the voids between the stones with smaller stones. The problem with this method is that the perfect puzzle pieces never seem to exist in nature, so I have to make a few of them.

To create a missing piece, I lay a piece of aluminum foil over the void and I fold back the edges until it fits the vacancy. I can also add bits of foil to the template if it's too small. When I've got it right, I carry the template to the pile of material and find the flagstone that most closely follows the lines of the foil. Then I trace around it with a nail, and saw off the excess stone. This method also works well for finding missing stones in a rubble wall. The foil can be used over and over again, and when you're done with the wall you can use the foil to bake a potato.

—Stephen M. Kennedy, Orrtanna, Pa.

Tuck pointing

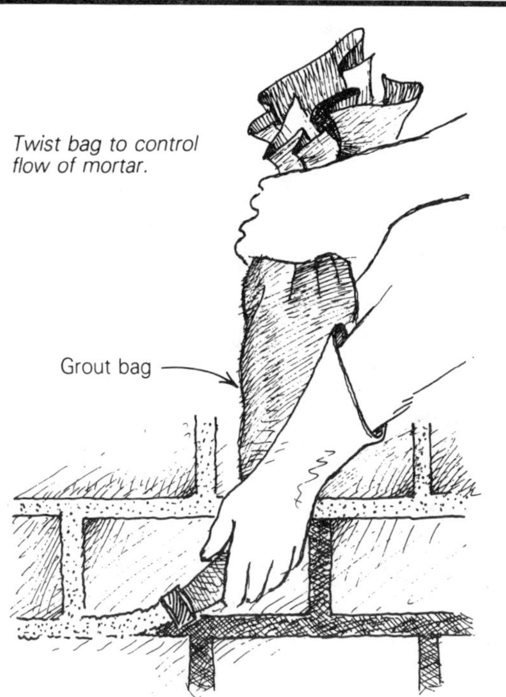

Nothing will ever make repointing old masonry a pleasure, but there is a tool that will take some of the drudgery out of it. Clean and prepare your joint lines as you usually do, then go out and buy a grout bag at a masonry supply store. I like the 24-in. model, but the size you'll wind up using depends on how strong you are—the bag gets heavy fast.

To use the bag, load in a mortar mix with just a touch more liquid than you normally use. Twist the open end of the bag or make a 90° fold in it to keep the mortar from oozing out the wrong way. Now position the nozzle so that the stream of mortar flows into your joint lines, as shown in the drawing. Wear rubber gloves during this operation, or plan on keeping the outside of the bag dry as a bone. If you don't, you'll wind up with peeling skin on both hands. Once you've got the joint lines loaded with mortar, go back and compact them with your striking tool.

—John Dobrin, Washington, D. C.

Concrete splint

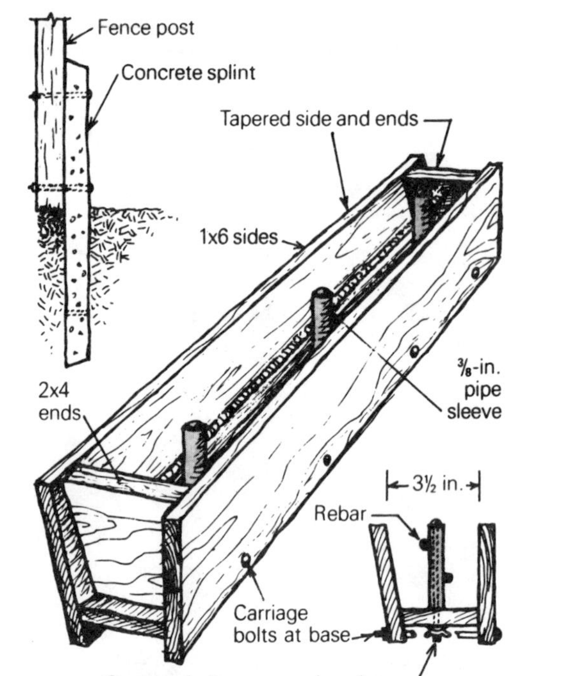

Carriage bolt secures pipe sleeves.

On a trip to England years ago, I saw an old fence whose posts had rotted off at ground level. The rest of the fence was still in good shape, so its owner had replaced the below-grade post sections with concrete splints. I've since mended my fences using the same solution, with excellent results.

I built a form about 4 ft. long, with sloping ends and one sloping side, as shown in the drawing. For each splint, three short pieces of ⅜-in. glavanized pipe are mounted on the form's axis to secure two lengths of ⅜-in. rebar. Two of the pipe sections are in the above-grade half of the splint, and act as sleeves for 5⁄16-in. bolts during installation.

Before filling the form, I lubricate the sides with a little axle grease to aid in removal. Nearly every splint I've cast has been with the left-over concrete from some small household job. *—Sidney McGaw, Albany, Calif.*

Foundation advice

I have saved a lot of money and headaches on a variety of new houses with this very simple technique: When I get bids for the foundation work, I insist on one combined price, including both the excavation and the concrete. This way each concrete subcontractor gets a price from his favorite excavator, and submits one complete price. What this means to the general contractor is that you won't find yourself faced with additional charges by the form setters for extra handwork on a hole that wasn't to their liking. In my experience, this includes most holes ever dug.

Besides eliminating surprise extras after the competitive relationship is lost, this strategy also results in one less subcontract you will have to let, schedule, supervise and shuffle paper for. Perhaps the greatest advantage, however, lies in simply avoiding a conflict between two subcontractors, with you in the middle.

—Douglass Ferrell, Trout Creek, Mont.

Anchoring to concrete

Anchoring wrought iron railings to concrete porches and stucco or concrete block walls can be a real problem. Pourable expansion cement offers a simple, reliable solution. It sets to usable hardness in half an hour and has twice the compressive strength of ordinary concrete. Expansion cement comes in powder form and can be mixed either to the consistency of putty for vertical work, or to pouring consistency for work on horizontal surfaces.

If you are working in old concrete, use a masonry drill or roto-hammer bit ½ in. larger than the diameter or diagonal measurement of the post to be installed, and drill a hole at least 2 in. deep. Rock and rotate the drill slightly to enlarge the bottom of the hole, then clean it out thoroughly, using an air compressor, bicycle pump or vacuum cleaner. Dampen the hole with water, set the post in place, and fill it with expansion cement. *—Geoff Alexander, Berkeley, Calif.*

Blocked-up chimney

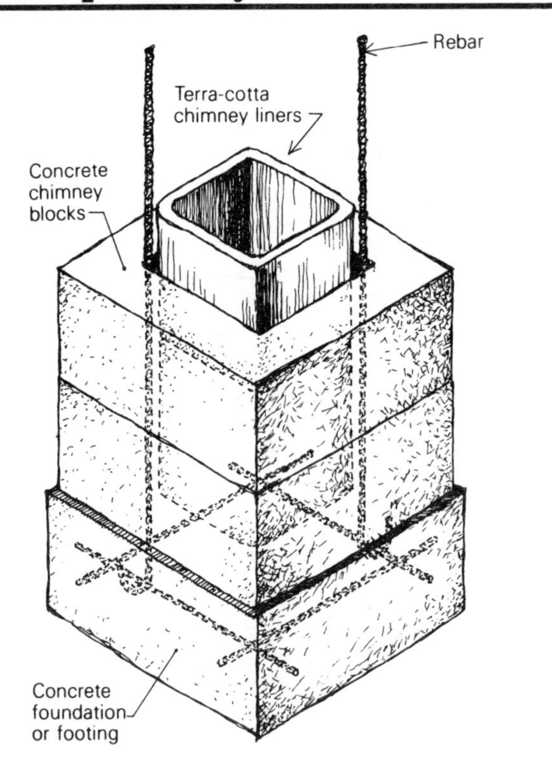

Prefabricated metal chimneys (whether insulated or not) have become almost universal for use with woodburning appliances. When we installed one, we thought, like most people, that our $20-a-foot expenditure was a lifetime investment. It hasn't been for us—after two years we have already had to paint the exposed pipe, which was badly corroded (in an admittedly severe coastal environment). We have heard that metal chimneys can be badly damaged by undetected chimney fires. With many builders installing them in hard-to-reach places, it seems that the alternative of terra-cotta chimney liners set in standard 16-in. by 16-in. by 8-in. concrete chimney blocks is a safer and more permanent choice. The cost is comparable or less, the units self-aligning, and the blocks can be tied together with rebar for earthquake resistance.

Flue liners and chimney blocks vary in size from location to location, but the local sources probably make compatible blocks and liners. The blocks around here are as shown in the drawing above. Liners are usually started 4 in. above the bottom so they align each chimney block as it is laid.

Depending on local codes, earthquake and wind conditions, and the height of the chimney, it's usually wise to run ½-in. rebars up from the footings to the top of the chimney on opposite diagonal corners and grout them in place. Usually 4-ft. to 5-ft. rods are as long as is comfortable to work with. Be sure to overlap them 20 diameters (10 in.). It is also advisable to tie the chimney to the ceiling and roof structure with steel straps, leaving at least a 2-in. clearance to combustible materials. —***Tom Bender, Nahalem, Ore.***

Gypboard concrete forms

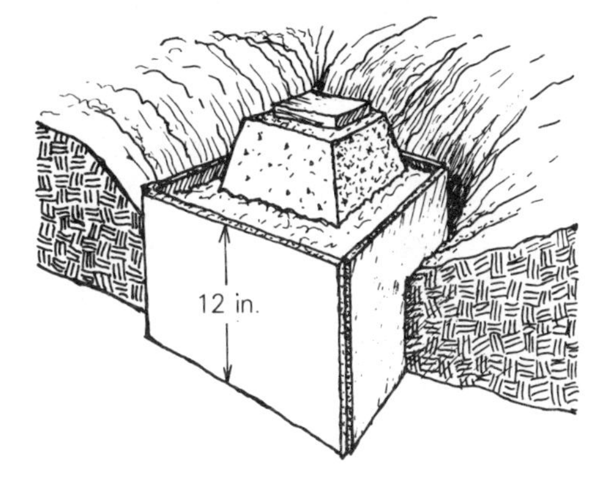

We recently did a foundation job in very crumbly, sandy soil. The first task was to set 27 pier blocks in pier holes that were 18 in. on a side and 12 in. deep. But by the time we had dug down a foot, we often had a hole that was more than 2 ft. across at the top and growing. Faced with filling these craters with concrete, we calculated that we would waste more than a cubic yard.

Instead of ordering the extra concrete, we transformed four sheets of gypboard into form boxes. We cut the sheets into 6-ft. lengths, and scored them along their length at 18 in. o.c., leaving the face paper intact. Perpendicular to these scored lines, we cut the board into 12-in. wide strips. These strips were then folded into square boxes, placed in the oversized holes and backfilled. The forms not only saved concrete, but also gave us an accurate way to calculate our ready-mix order. —***Sunrise Builders, Santa Cruz, Calif.***

Brick-path screed

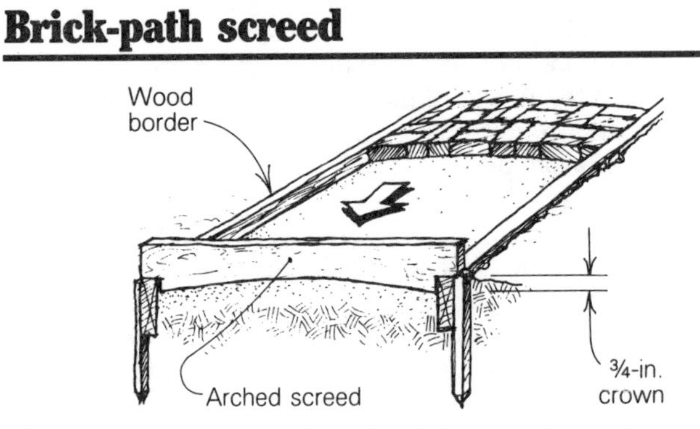

All walks, drives and patios should be constructed to ensure proper drainage. Concrete sidewalks are normally flat, but walks made of brick, tile or paving blocks should have a slight crown built into them. The crown promotes drainage, minimizing the effects of the freeze-thaw cycle, and eliminates puddling.

Before laying a brick path, I use a wood screed with a slight arch to contour the sand bed. The screed has a notch cut into each end, as shown above. The notches, which are ¼ in. shallower than the thickness of the brick I'm using, ride on the path's wood borders. Before I use the screed, I moisten and tamp the sand to minimize settling.

—Rod Goettelmann, Vincentown, N. J.

Through-wall pipe forms

Couplings at both ends

Forms

Running pipe or conduit through a poured concrete foundation wall with no gaps to patch and without cutting holes in the forms sounds too good to be true until you've tried this trick. Cut a piece of the pipe you need and attach couplings to both ends so that the total length of pipe plus couplings exactly matches the thickness of your wall, as shown in the drawing.

This unit will then fit snugly between the forms, and can be wired to the rebar and spreaders at the top of the forms to keep it in place during the pour. When the forms are pulled away, you have a coupling embedded on each side, flush with the wall and ready for another length of pipe.

—D. A. Fleury, Curlew, Wash.

PVC-clad piers

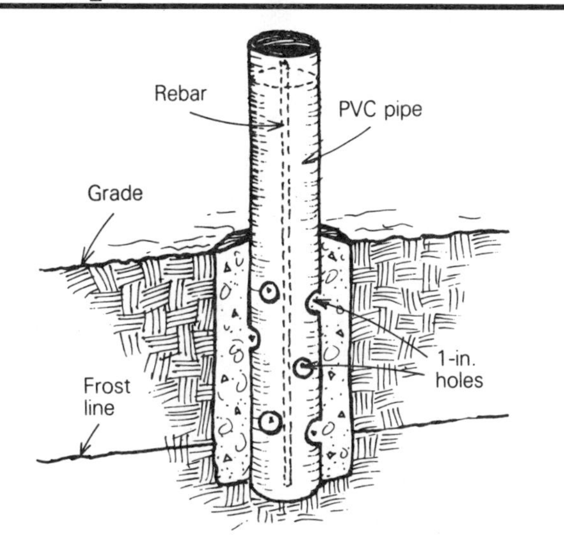

The drawing above shows my pier design for supporting decks and small buildings. The key to the system is the section of PVC drainpipe, which acts as both form and finish surface. I like the piers produced with this method because they are strong, cheap, easy to install, rustproof, rotproof and good-looking.

I begin a pier by digging a 6-in. to 8-in. dia. hole with a post-hole auger to the required depth below the frost line. Next I use an electric drill with a wood-boring bit to punch several holes in a length of 4-in. dia. PVC pipe. The holes allow the concrete inside and outside the pipe to combine for strength. For heavy or high structures I use 6-in. pipe.

Then I set the PVC pipe in the hole, leaving it a few inches longer than the final cut-off height. I pour concrete into the hole to grade level, and into the pipe to the desired height—a foot or so above grade. Then I set ½-in. rebar into the center of the pipe. The rebar can be left long to act as a pin, or be pushed below the level of the concrete. A J-bolt, angle iron or other anchoring device can be added as well. While the concrete is still wet, I plumb the pipe and let the pier set overnight.

I made a cut-off marking jig from a 4-in. plastic coupling by grinding off the interior ridge so that the complete coupling slides down over the pipe. I mark the pipe for finished height with a scribe held against the jig, and cut it off with a reciprocating saw. Finally, I fill the remaining section of pipe with concrete and trowel it flush with the top of the pipe.

—Robert A. Ritchie, Westerly, R. I.

Corrugated column forms

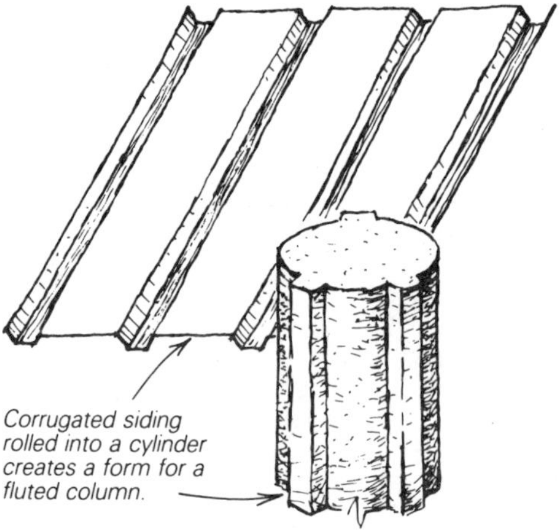

Corrugated siding rolled into a cylinder creates a form for a fluted column.

Recently I bid on a job that called for 15 concrete piers, each 30 in. high. The loads involved weren't much—about 100 lb. per pier, so the job didn't require the bearing or expense of a standard 10-in. column. In addition, large-diameter columns would have appeared too massive for the job.

I used corrugated metal siding to form these short columns. It's available in various lengths, widths and gauges. The siding I used has 1-in. deep corrugations that are 2 in. wide, spaced 10 in. apart. When I rolled up a section of the siding, it formed a column about 8 in. in diameter with three 2-in. flutes. I held these sections fast by securing the overlapping metal with three short sheet-metal screws.

I made five forms out of this material, and made three pours with them. The siding's slick finish made for easy release from the concrete, and an attractive, smooth finish.

—E. L. Doonan, Aledo, Ill.

Epoxy syringe

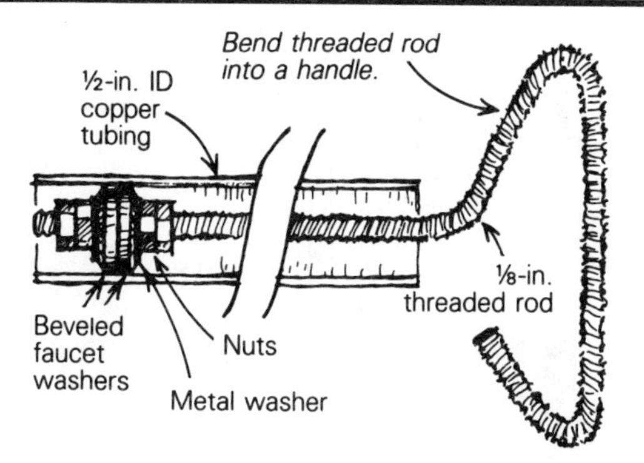

My crew and I have been adding a second story to an older home, but before we could get to the framing we had to deal with the foundation. The original footing was too narrow to carry the weight of the addition, so we had to widen it by adding new footings alongside the old ones. To connect the old and new concrete, the engineer called for ¾-in. rebar dowels. And the dowels had to be epoxied into the old footings. I checked out the epoxy in glass capsules that are made for this purpose, but at around $4 apiece they were more than we wanted to spend—we had hundreds of empty holes to fill.

My research led me to an outfit called Adhesive Engineering (1411 Industrial Rd., San Carlos, Calif. 94070), which makes an epoxy called Concresive 1411 that is intended for just this purpose. It is thick enough to stay in a horizontal hole without drooling out before it sets up. A batch of it cures slowly—about a two-hour pot life at 65°F. But once in the pot, how to get it into the holes?

Our solution is shown in the drawing above. It is a syringe made of a length of ½-in. ID copper tubing, with a plunger of ⅛-in. dia. threaded rod. At one end of the rod, I made a piston out of a pair of beveled faucet washers. They are held fast by washers and nuts. Tightening the nuts increases the diameter of the piston a bit, ensuring a good fit. To fill the syringe with a dose of epoxy, put the tube in the pot and pull up on the plunger. Now you're ready to expel the adhesive into the target hole. Using this method, we bonded all the required dowels at a cost of about 75 cents apiece.

—Joe Wilkinson, Berkeley, Calif.

Contour gauge for piers

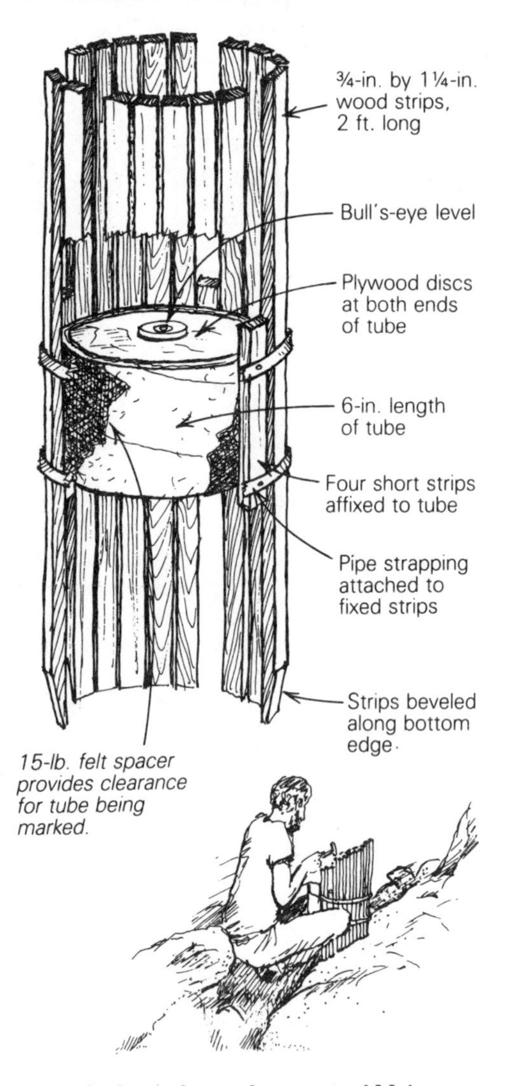

Constructing a deck on the rocky coast of Maine presents the problem of accurately and quickly cutting tube forms so that they conform to exposed, irregular bedrock surfaces. Good-looking, strong concrete piers require a leak-free contact between the form and the rock. To meet these requirements, I constructed the contour gauge shown in the drawing. Pressed against the rock and made plumb with the aid of its bull's-eye level, the slats of the gauge can be tapped to conform to the contour of the rock, and this line can be quickly transferred by inserting the tube to be cut into the gauge, and tracing around the bottom of the movable strips. I built this gauge from scrap materials in a couple of hours, and it easily paid for itself in one job.

—Ian R. Walker, Princeton, N. J.

Form stops

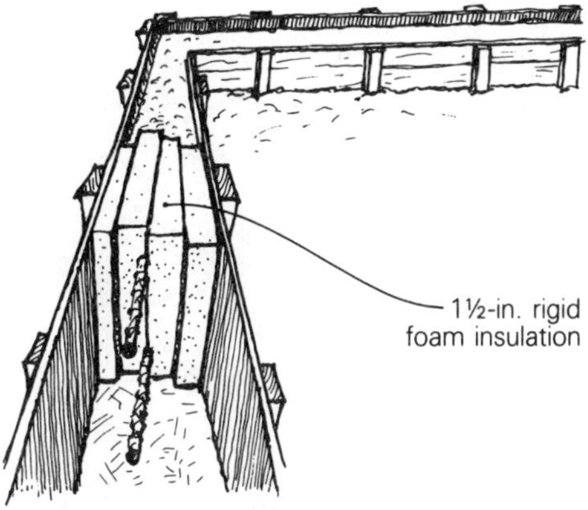

1½-in. rigid foam insulation

You can't always avoid cold joints in concrete, particularly in renovation work where things have to proceed incrementally. Here's an easy way to make a quick dam in your forms, even in places with lousy access.

Wedge pieces of 1½-in. rigid insulation into the forms like a row of books, as shown in the drawing above. It's easy to cut channels in the foam for the protruding rebar, and the pieces will come out fairly easily once the pour has set up. Make sure to stagger the edges of the foam pieces to form a keyway for the next pour. *—Jim Ramsay, Kelowna, B. C.*

Dry-wall advice

When I build or repair a mortarless stone wall, I fill the voids between the stones with gravel. This makes the wall more stable by locking the larger stones in place. It will also increase the mass of the wall—a wall built without mortar can be nearly half air space.

If you don't "dry-grout" like this, nature has a way of

gradually filling the voids with soil, roots, vines, mice and snakes. This organic material can often be the downfall of a dry wall. The gravel will also help to prevent the damage caused by frost heaving.

—Stephen M. Kennedy, Orrtanna, Pa.

Form bracing in loose soil

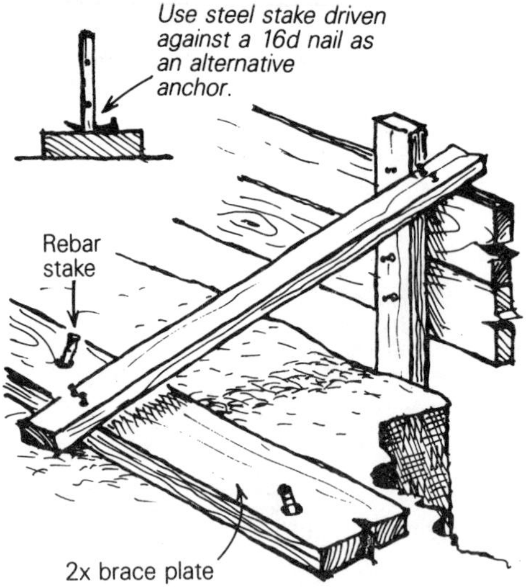

It can be frustrating to brace concrete forms in loose, sandy soil—especially when the stakes begin to creep outward as the forms are filled. Faced with these soil conditions, I recently used a 2x brace plate to anchor my form braces.

As shown in the drawing above, I positioned the plate far enough from my forms to give my braces about a 1:2 slope. The plate has 1-in. holes drilled 2 ft. on center. Through these holes I drove 2-ft. lengths of #4 rebar, oriented at opposing angles. Secured this way, the plate served as a sturdy anchor for my 2x4 braces.

If you've got some steel stakes, use them to anchor the plate. As the end of the stake draws near to the plate, insert a 16d nail halfway through one of the holes and drive the stake a little farther until the nail begins to bend as it engages the wood. When it comes time to remove the plates, lift them out of the ground with a backhoe. Lacking the backhoe, a pair of locking pliers makes a good handle on each stake.

—Michael Hermann, Nevada City, Calif.

Brick cutter

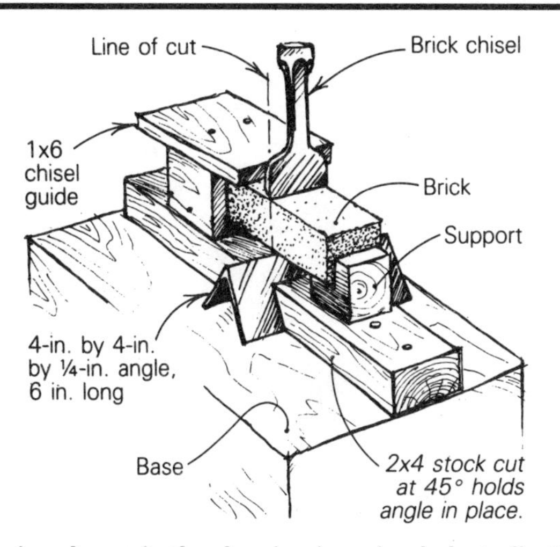

The device shown in the drawing is a simple but effective brick cutter that works by shearing a brick between a fixed angle iron and a brick chisel. It's not as fast as breaking bricks with a mason's hammer, but there will be more accurate cuts and a lot less waste. To make the cutter, file a true edge on the outside corner of a short piece of angle iron and place it on a heavy base, such as a beam offcut, with the oustide corner facing up. Secure the angle by placing the mitered end of a 2x tight against each side.

On one side of the angle, position a 1x6 up on a thick block to act as as guide for the brick chisel. Be sure to set the guide high enough to clear the thickest brick you plan to cut. Adjust its length so that when the bevel side of the brick chisel is held tight against the guide, the point of the chisel is directly over the edge of the angle iron. On the opposite side of the angle, place a support block to cradle the brick.

To use the cutter, place the brick on top of the angle with your mark centered over its edge. Position the chisel on top of the brick, bevel side tight against the guide. One or two blows with a heavy hammer should do the job. For face brick, cut the brick ⅛ in. to the waste side of the mark and clean up the exposed edge with short, controlled paring strokes of the chisel.

—Will Foster, Aberdeen, Wash.

Corrugated concrete chute

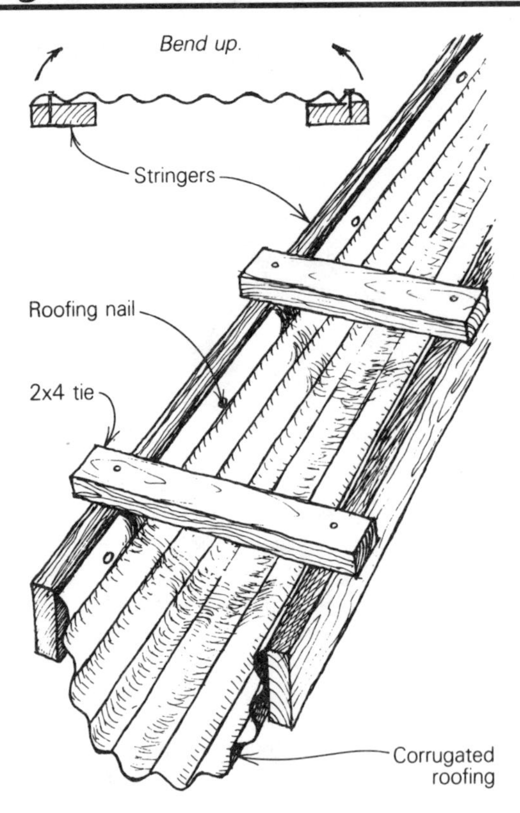

When the ready-mix truck runs out of chute, it's easy to extend your reach with the site-built job shown above. It's made of corrugated steel roofing that has been nailed to two flat 2x6 stringers and bent into a U shape. It maintains its profile with the help of 2x4 ties nailed across the top, and it is secured to the truck chute with a length of chain. Wet concrete is mighty heavy, and I consider 2x6 stringers to be the absolute minimum. My 16-ft. chute is strung on 2x10s, but it's somewhat overbuilt.

Without the 2x4 ties, the chute can be usedas a slide for concrete blocks, as the corrugations eliminate much of the drag. Sure beats using a wheelbarrow.

—Al Dorsa, St. Croix, V. I.

Concrete calculations

Figuring out a load of concrete for foundations, walls or slabs can seem like an endless math assignment. The following formulas are shortcuts to the conversions from square inches to cubic yards. For footings, multiply the cross section in square inches by .000257 to get the cubic yards in one lineal foot. Multiply that result by the length of the footing for your total. For example: Find the number of cubic yards of concrete in a footing 14 in. wide, 17 in. deep and 43 ft. long.

$14 \times 17 \times .000257 \times 43 = 2.63$ cu. yd.

This formula also works for columns. Just think of them as vertical footings.

For walls and slabs, it's a little different. Multiply the slab or wall area in square feet times the thickness in inches times .0031. For example: How many cubic yards of concrete are there in a wall 4 ft. high, 82 ft. long and9 in. thick?

4 ft. $\times$ 82 ft. $\times$ 9 in. $\times$ .0031 = 9.15 cu. yd.

—Charles Fockaert, Eureka, Calif.

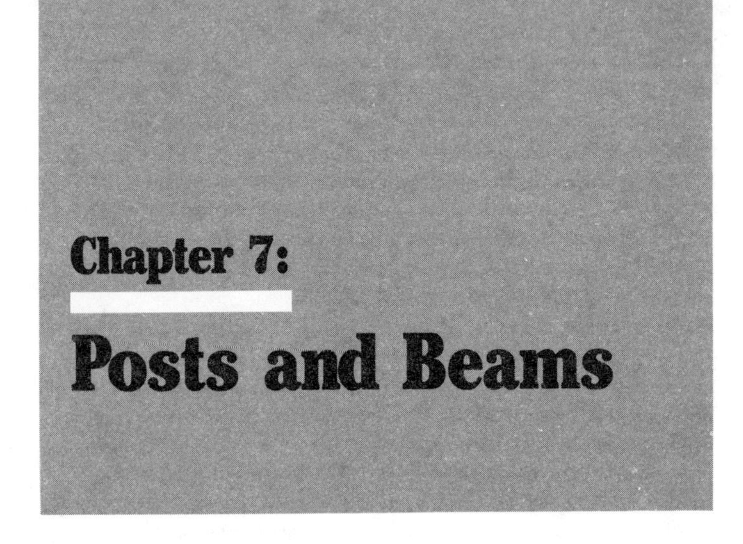

Chapter 7:

Posts and Beams

Cutting curves in big beams

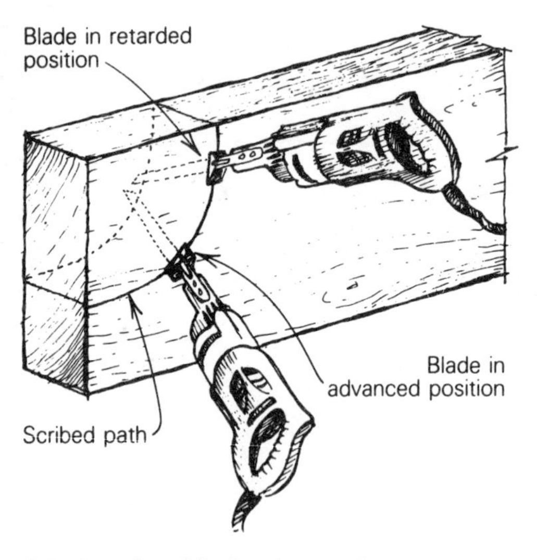

Last year I designed and built a house that uses seven cantilevered 6x20 beams to hold up the second floor. The beams are exposed to view on the outside, so I wanted a decorative cut on each one to dress it up. The simple 12-in.

radius arc I chose turned out to be a lot easier to draw than to cut with the tool at hand—a Sawzall with a 9-in. blade.

The basic problem with cutting curves in thick stock with a reciprocating saw is that the free end of the blade tends to drift outward, cutting an arc larger than the one that is being followed. After much trial and error, including guiding the protruding blade with a pair of lineman's pliers, I developed the technique explained below.

Begin by scribing each beam carefully on both sides with the aid of a cardboard template, and then place it upside down on a pair of sawhorses. Once approximately ¼ in. of the cut has been made, the operator can advance or retard the blade on the scribed path by leading the cut with either the heel or the tip of the blade. This develops a torsion within the blade that affects its course.

Have a helper stand on the opposite side of the beam and describe the path of the blade as the arc is being cut. If the blade moves outside the pencil mark, the operator needs to advance the blade. If the blade starts to move inside the line, the operator can retard it to pull it back on course. With a little practice, this method works very well.

Incidentally, the helper is actually relief personnel, since each cut on my 6x20 beams took nearly 1½ hours to complete.

—Eric K. Rekdahl, Berkeley, Calif.

Timber moving

If your building project calls for moving large timbers over rough ground, lay your hands on some 55-gal. oil drums. They make great rollers on uneven surfaces.

—Jim Sergent, Bainbridge Island, Wash.

Easy-tie lumber flag

Whenever I have to carry lumber that extends beyond the bed of my pickup, I just tie on my Velcro flag. I knot one corner of a red bandana around a strip of Velcro about 1 ft. long, and then tighten the Velcro around a projecting stick of lumber. This is easier than driving a nail into the lumber or tying a string around it to hold the flag, and the Velcro hangs on tight—even at highway speeds.

—Tom Richards, Dorset, Vt.

Rolling bandsaw

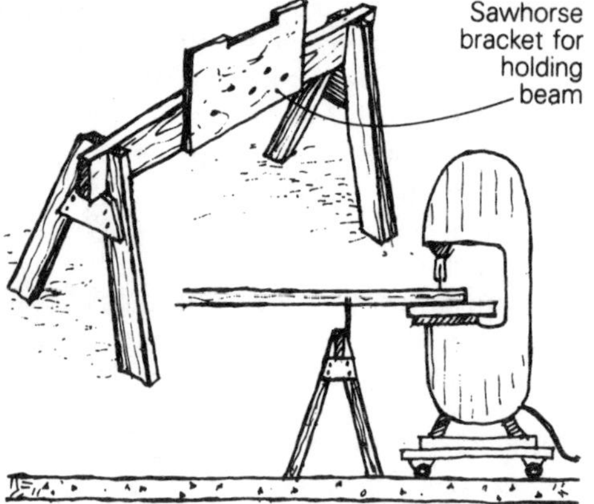

I have an alternate method for cutting curves in thick timber with a Sawzall. It was devised after rejecting several other approaches, including the portable bandsaw, which had insufficient throat for the cuts we wanted to make.

I had to make decorative curved cuts in large beams. I trucked the beams to my shop and set up my bandsaw on a dolly mounted on casters. Then I cut a pair of plywood supports with shallow notches the width of the beams, and spiked them to the 2x6 crossbars on a pair of sturdy sawhorses as shown above. The bottoms of the notches were placed at the height of the saw table, and the horses were spread apart to support the beams securely.

After marking, the beam was placed on the horses, and the bandsaw was rolled up to the beam to make the cut. Since the dolly could readily be moved in any direction, I found that I could direct the blade accurately through the wood, albeit using a spread-legged stance and bear-hugging the saw.

The first wheels that I used were the ball-bearing type that are set in their own chase. But these tended to collect sawdust, causing the balls to stop rolling. I ended up using offset rubber wheels, which worked much better.

With a ⅜-in. blade, it was easy to cut the 5½-in. diameter semicircle I had drawn on the beam. I had an assistant along that morning, but the setup was so stable that he had little to do beyond catching the cut pieces as I rolled the bandsaw around the beam.

—Robert Gay, Seattle, Wash.

Nested post-to-beam joint

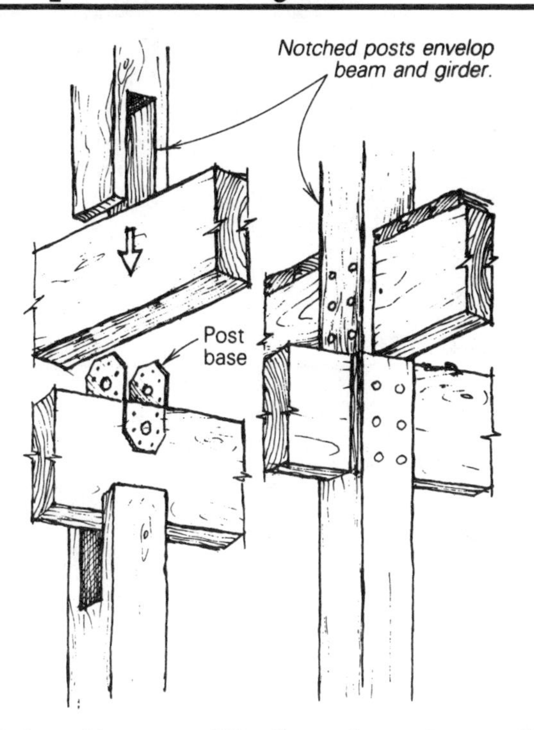

The wind can blow up to 100 miles an hour where we live on the Oregon coast, so it's mighty important for a roof to be tied firmly to its foundation. To hold ours down, I needed a tension tie between a post, a girder, a beam and another post upstairs. There are lots of steel brackets that can make these connections, but I don't like looking at them. So I devised the joint shown above.

First I nailed a Simpson (1450 Doolittle Dr., San Leandro, Calif. 94577) B6 post base to the top of my girder, and then I lowered it into a notch in a post sized 2 in. wider than the girder. The upper flanges of the post base tie the beam to the girder, and I notched the upstairs post to slip over the beam. This conceals the hardware and gives the appearance of a continuous post with beams passing through it. I used hardwood pegs to tie the beam and girder to the posts.

—Tom Bender, Nehalem, Ore.

Plywood protection

If your job-site plywood stack seems to be shrinking when you're not around, try screwing it down. I use 3-in. drywall screws to connect the top layers at the corners. The resulting 3-in. thick plywood slab is heavy enough to discourage all but the most determined pilferers, and your plywood and whatever lies under it will stay put until needed.

—Michael R. Sweem, Downey, Calif.

Built-up beam

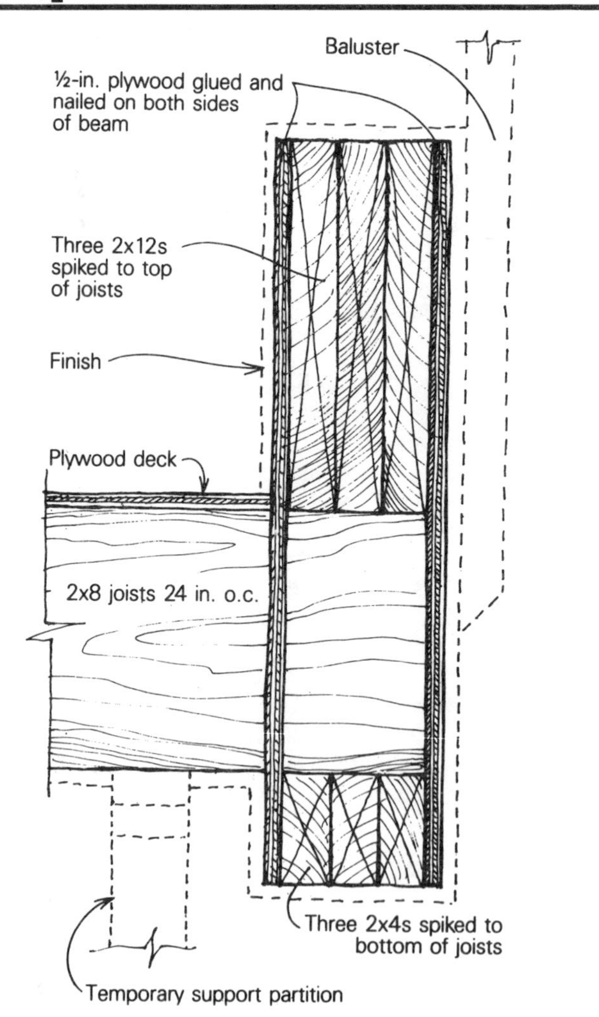

The house I recently worked on has a balcony that spans 23 ft. over the living room. The plans called for a 6-in. by 22-in. by 24-ft. beam to carry the balcony, but such a beam proved hard to find—even a laminated version. We considered using a steel I-beam, but the rough framing was nearly complete, and a soggy site suggested a difficult delivery. So builder Charlie Callahan and I worked out this simple site-built plywood box beam as an alternative.

First we set the balcony floor joists on a temporary partition, as shown in the drawing on the facing page. Then we spiked together three 2x12s on top of the joists, forming a balcony curb. Three 2x4s were spiked together under the joists for the bottom chord of the beam. Since our clear span was 23 ft. and we needed 6 in. of bearing at both ends, we cast about for 24-ft. framing lumber. We were fortunate to find some locally, so we didn't need to splice any chords. However, splices are permissible in this kind of beam if they are staggered, and glued as well as nailed.

To connect the top and bottom chords of the beam, we covered both sides with ½-in. plywood, nailed and glued. We used sheets 22½ in. by 8 ft. on the outside face of the beam. On the inside face, we covered the spaces between the joists with pieces that were 22½ in. square. The beam is wrapped with T&G oak as a finish.

This beam is exceptionally strong, and it acts as one with the balcony joists to resist twisting. It was also fast and easy to build in place. We engineered the beam with the help of American Plywood Association's "Plywood Design Specification," Supplement #2. Single copies are available free from APA, 7011 S. 19th St., Tacoma, Wash., 98466.

—Frank Lee, Baltimore, Md.

Moving lumber

We often do large jobs far from the street, which makes lumber delivery by truck impossible. In these situations, I arrange for the lumber delivery on the day we plan to backfill. This way the lumber company drops the banded lumber alongside the street, and we have the bulldozer operator move the lumber. He can either carry the lumber in the bucket or use a chain sling. This can save you a good bit of labor.

—Patrick D. Rabbitt, Greenwich, Conn.

Beam boom

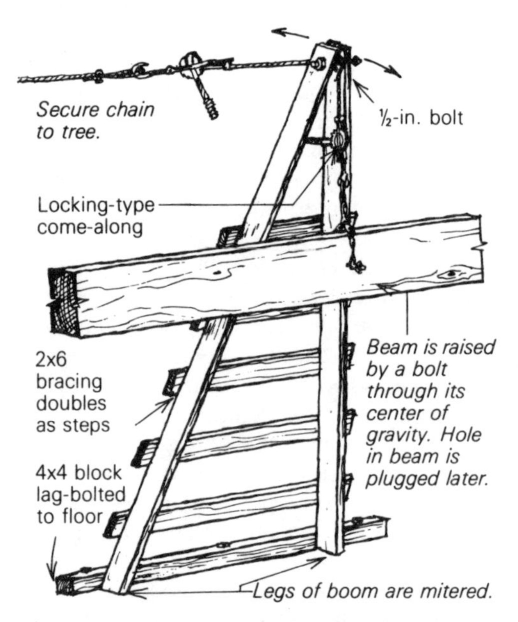

An old-timer showed me this device for lifting heavy timbers single-handedly. I have used it to raise 22-ft. long 4-in. by 16-in. laminated beams with surprising ease. The device requires two blocks and tackles or ratchet come-alongs with locking devices, a boom, various lengths of chain or cable, miscellaneous hardware, and a tree or equally substantial mooring.

Construct a boom as shown in the drawing. It should be long enough to be at roughly 60° to the floor when the beam is lifted to its finished height. Lag-bolt a heavy block (at least a 4x4) to the floor to prevent the bottom of the boom from moving and position the boom over the timber at its center point by adjusting the two come-alongs. Drill a bolt hole at the balance point in the beam, attach it to the vertical come-along, and lift away. Control over the arc of the boom and the elevation of the beam is easily adjusted with the two come-alongs, allowing precise horizontal and vertical positioning.

After raising a beam to the desired height, I add a second chain wrapped around the top of the boom and attach it to the tree as a safety measure; it has proved to be unnecessary, but makes me feel more secure while the beam is aligned and pegged into place.

—Robert VonDrachek, Missoula, Mont.

Router-bit drip edges

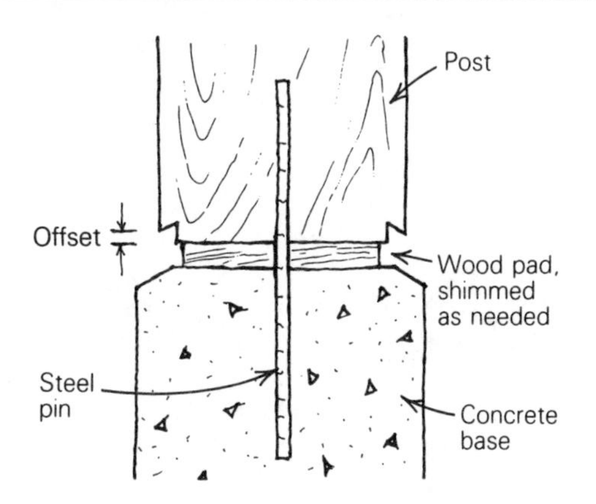

You can use a dovetail router bit to shape a drip edge that's just right for all sorts of situations. For example, to protect a square post base against end-grain rot I used one of these bits to cut the drip edge shown in the drawing above. The angle cut by a dovetail bit is steep enough to shed water, but shallow enough to be strong. When I cut a drip edge on a post, I allow some offset between the bottom of the post and the drip edge. This allows me to insert any shims without dinging the edge.

—Emanuel Jannasch, Chester, N. S.

Chapter 8: Frame Carpentry

Raising walls

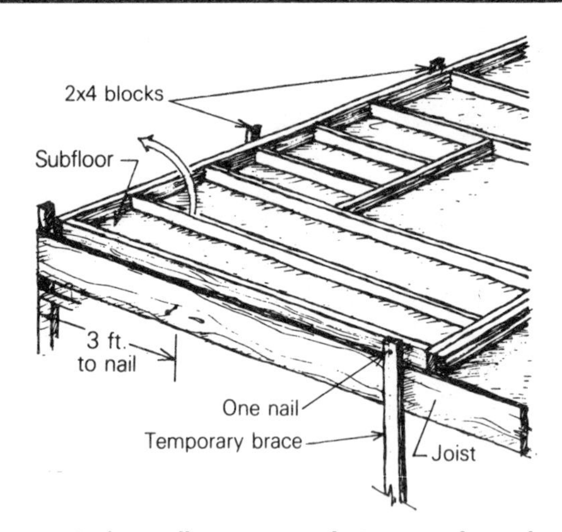

Raising an exterior wall on a second story can be awkward. Here are two tips to make this operation safer. Nail 12-in. 2x4s every 4 ft. to the outside edge of the platform where the wall is to be placed, leaving about 4 in. of this block sticking above the platform. When you raise the wall, the stops will prevent the bottom plate from sliding off the subfloor.

To eliminate frantic bracing after the wall has been raised, nail braces near the top of each end of the wall with only one nail per brace. If the wall is on an exterior corner, drive a nail into the joist or blocking about 3 ft. from the corner, as shown in the drawing. Then when the wall is raised, the brace will swing freely and come to rest on the nail, where it can be attached quickly as a temporary support. *—Don Dunkley, Sacramento, Calif.*

Stud-wall adjustments

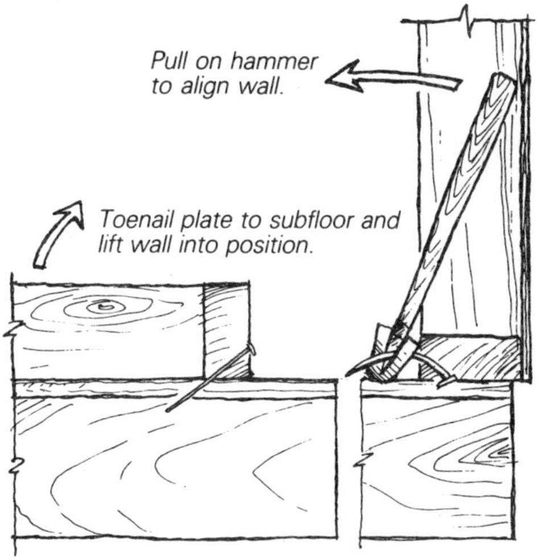

Pull on hammer to align wall.

Toenail plate to subfloor and lift wall into position.

I usually put the plywood sheathing on my exterior stud walls before I raise them into position. However, once the wall has been raised it often takes four hands to move it around, and it can be quite awkward to maneuver the bottom plate toward the layout line. Here's a clever way to handle this problem that I learned from some production carpenters.

While the wall is lying on the subfloor and before the plywood is applied, toenail the bottom plate to its layout line with 16d nails, as shown in the drawing above. As the wall is raised, these nails will bend and lift out of the floor. These are your handles. Using your hammer as a lever, you can pull the wall inward to align it. Nail it home, then nip off the bent nails with your side cutters.

—Tom Strong, Westcliffe, Colo.

Framing jig

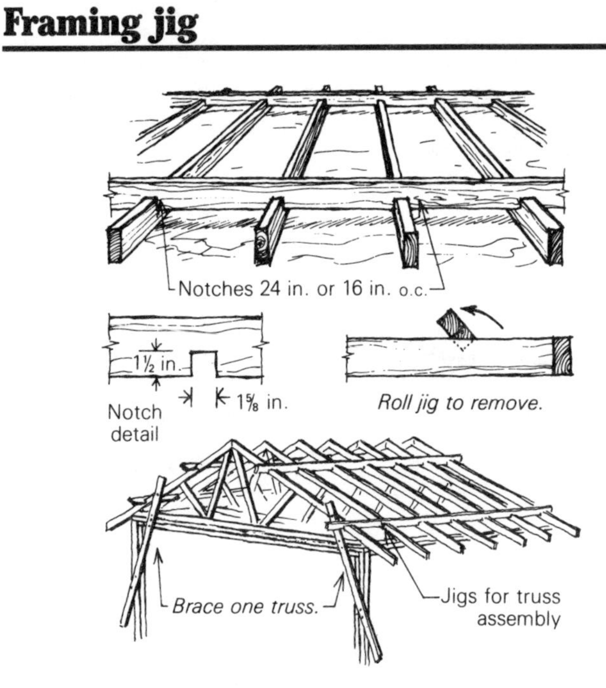

I have a rapid and accurate method for assembling any wall, floor or roof that uses standard dimensions. It consists of a jig, which is simply a 2x4 notched on 16-in. or 24-in. centerlines to receive the framing members. Two such jigs are handy, and for large-scale work, four can help. To use, just slip one member at a time into the jig until everything is in place, then nail. No measuring is required except for cutting pieces to length. It really helps on long runs that need to be sheetrocked, covered with plywood, etc. For trusses, it means that everything can get assembled via the jigs, the plywood started and then the jigs pulled up. No nails to pull.

I put together a 1,000-sq. ft. workshop with this method—mostly by myself and on evenings and weekends. It really works. ***—M. R. Havens, St. Albans, W. Va.***

A strapping idea

The easiest way to keep the bottom plate of a framed wall from walking when it is being raised is to toenail it into the subfloor. The nails bend easily as you lift the wall, and the

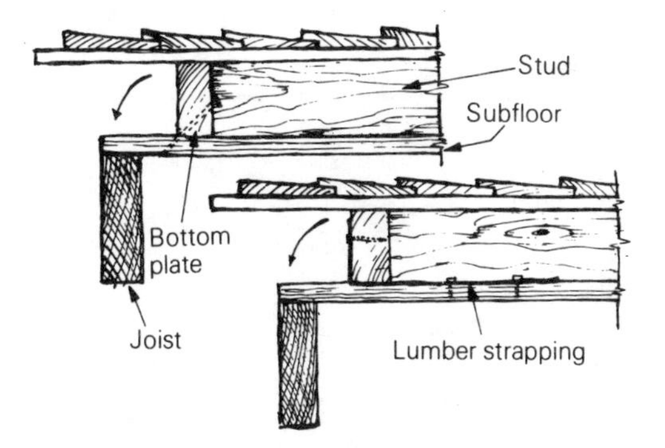

bottom plate usually remains in the immediate neighborhood of its intended layout.

A more secure system uses the strapping that binds lumber loads. Cut it into 12-in. pieces and nail one end to the underside of the bottom plate. The other end should run under the wall and will be nailed into the subfloor. Concrete nails will pierce the stuff, or you can abuse your 2/32 nailset and start a hole. The strapping can be left in place and covered by the finish flooring. ***—M. F. Marti, Monroe, Ore.***

Separating studs

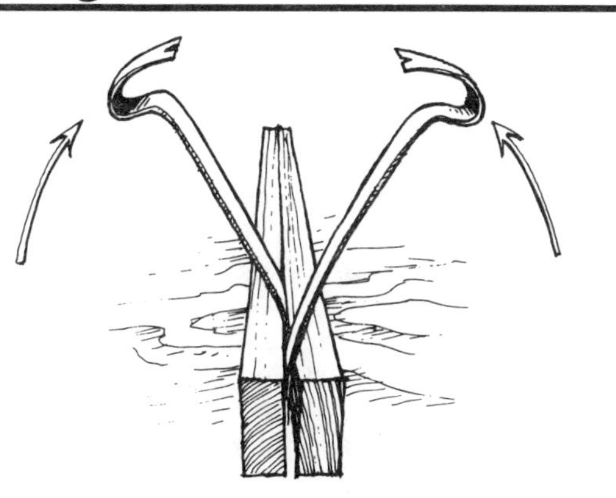

Seeing me struggling with my framing hammer, trying to separate two studs nailed together, prompted the carpenter I was working for to show me this trick. Use two pry bars, one in each hand. Insert the flattened ends between the studs and simply close together, as shown, working your way down the studs. Saves the back and clears the air.

—Gerald McSweeney, Toronto, Ont.

The hooter stick

Plumbing and aligning stud walls can be quite a chore, sometimes involving the better part of a five or six-man crew. On a recent job, I became acquainted with the tool shown in the drawing above. Here in Austin, Tex., it's called a hooter stick, and I haven't found anything that's better suited for adjusting long, tall or just plain awkward walls.

Basically it is nothing more than two studs, a 2x4 block 20 in. to 30 in. long and an old hinge. To assemble the stick, first cut a 45° V-notch in the end of one stud and scab the block flush to the bottom end of the other stud. Then fasten the two parts with the hinge.

To use the hooter stick, place the notched end against the underside of the top plate, near a corner or an intersection with another wall. To brace the bottom of the stick you can use either your foot or a block that is tacked to the subfloor. Now you're ready to push in the direction that you want the wall to move. The hooter is an awkward piece of equipment to manipulate at first, but once you get used to it you'll be surprised at what you can do to an outside wall full of offsets and headers. ***—Paul Wilson, Austin, Tex.***

Hand room

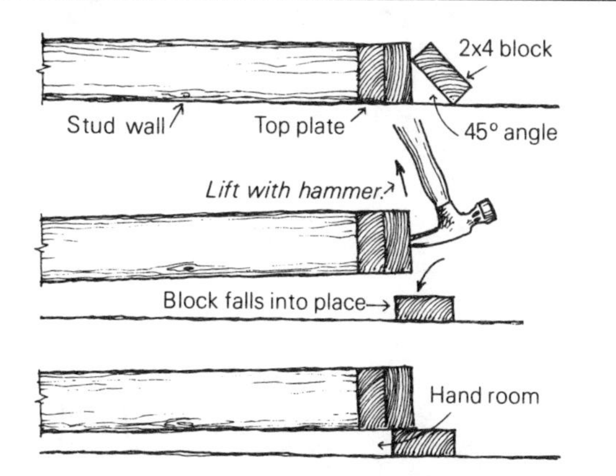

Back when I worked in the tracts, a little trick made our lives easier when it came time to lift stud walls into place. Both people working on a wall would pick up a scrap piece of 2x4 and lean it against the top plate at about a 45° angle. Then we would bury the claw end of our hammers into the top plate deep enough to provide lifting purchase on the wall. Both workers lifting together would then raise the entire wall enough to allow the 2x4 blocks to fall under the top plate, providing hand room for the final lift from the other side.

—Dave Bullen, Berkeley, Calif.

Copper caps for rafters

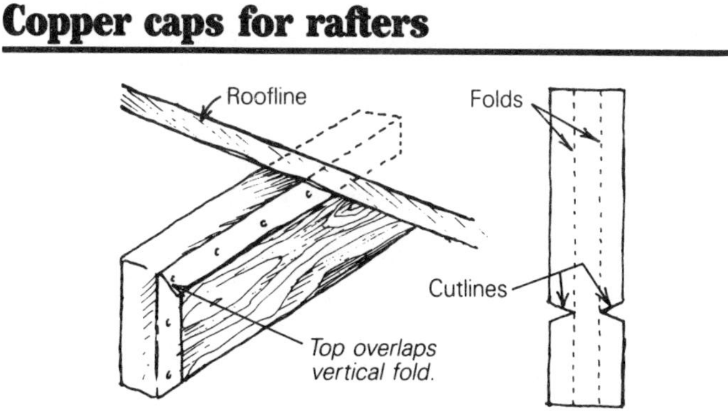

The ends of exposed rafters that extend beyond the roofline will decay if they're left unprotected. Paint or stain works for a while, but both have to be renewed from time to time.

Copper end caps are a good solution to this problem. I've used them on several jobs, and they are easy to install. They cost about the same as gutters, and look better with each passing year. I used 16-oz. copper sheeting, cut to the pattern shown and nailed in place with brass ringshank nails to avoid any galvanic action.

—Angelo Margolis, Sebastopol, Calif.

Walls with braces

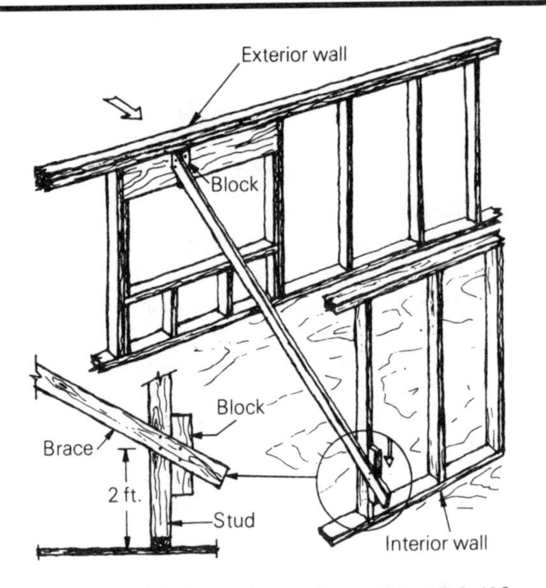

When plumbing and lining a frame, I use this trick if I encounter a stubborn exterior wall, bristling with headers and bowed outward.

Nail a 2x4 scrap flat to the outside of the exterior wall below the top plate where it is most bowed out. Secure a long 2x4 brace onto the side of the block. The brace must be long enough to reach an interior wall that runs parallel to the exterior wall you are lining. Hold the bottom end of the brace about 2 ft. off the floor and against a stud in the interior wall.

Attach an 8-in. block to the brace, making sure that the block is on the far side of the interior wall. Now push down on the brace at the point where the block is attached. The levering action of the block against the stud will pull in any bow in the exterior wall. Once the wall is straight, the brace can be face-nailed to the stud.

—Don Dunkley, Sacramento, Calif.

Framing under a peak

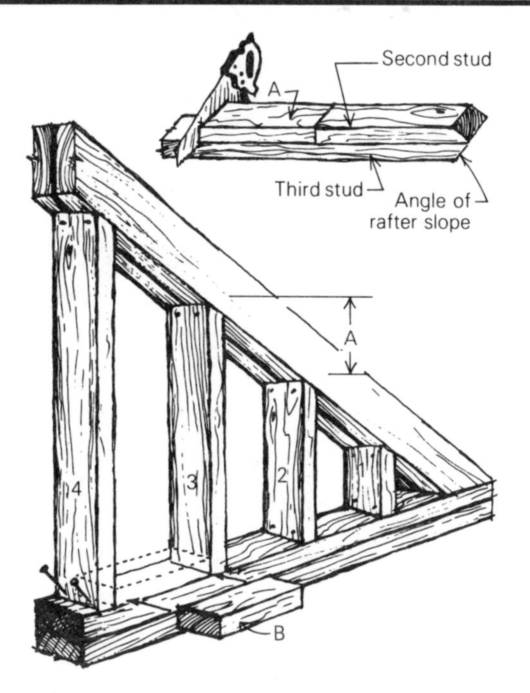

Putting studs into the wall beneath the end rafters of a peaked roof can be time-consuming. But two simple jigs make it easy. Using a carpenter's square (or a calculator), determine the change in stud length (A in the drawing above). Assuming that the studs are equally spaced, this dimension is constant the entire length of the rafter. Now cut the ends of the studs at the angle that fits the rafter slope. Next, cut a piece of stud scrap to length A. The length of the shortest stud is found by measuring; from then on add the scrap block to the length of the previous stud and mark a cutting line, as shown.

Once cut, the studs will nail up easily if you make one more block. Cut another scrap piece as long as the distance between the stud faces (B in the drawing), and use this as a spacer block. If you work accurately, the tops of the studs will wind up in the right position just by nailing.

—Kevin Kelly, Westfield, N. J.

Freestanding scaffolding

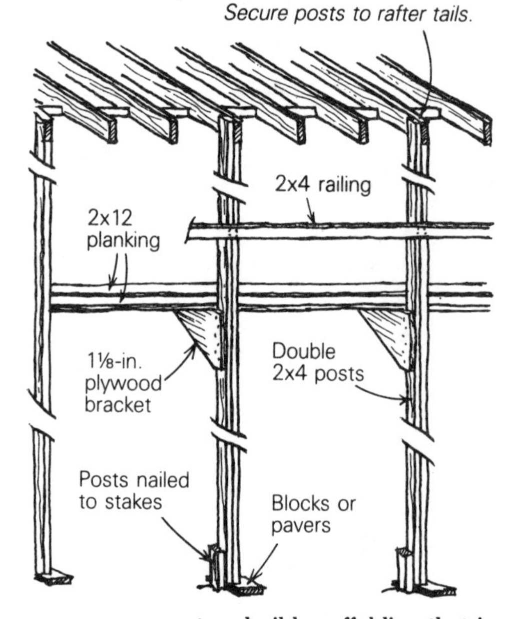

I have seen many carpenters build scaffolding that is attached to the sides of the structure, which means it is useless when it comes time to trim out and paint the building. The drawing above illustrates a simple design for a scaffold that doesn't rely on wall mounts. I use it when the structure has exposed rafter tails that project well beyond the walls.

The vertical supports for the scaffold are posts made of pairs of 2x4s. They are secured at the top by nailing them to the rafter tails. At the bottom, I stand the posts on wooden blocks or concrete pavers to keep them from settling and secure them to stakes driven into the ground. I install the posts 6 ft. to 8 ft. on center, depending on the type of planking I'm using.

For my scaffold brackets, I use 1⅛-in. thick plywood. I make the brackets big enough to hold two planks side by side. I attach them at the desired height with plenty of 16d nails. After I lay down the planking, I usually add a 2x4 railing for increased security.

—Michael Gornik, Nevada City, Calif.

Precut plates

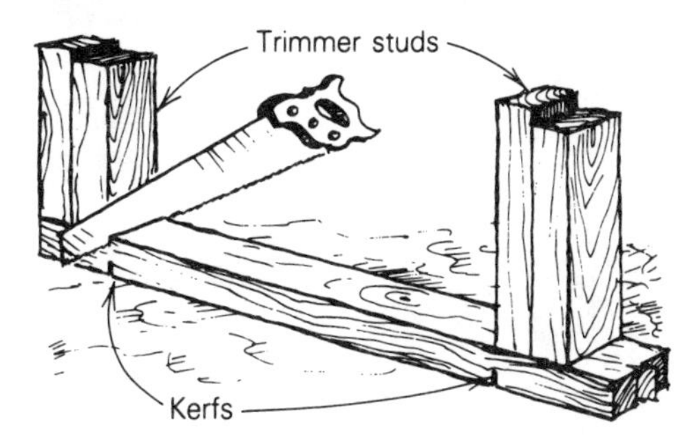

When I build stick-framed walls, I precut a portion of the sole plate where it has to be removed for doorways. I use my skillsaw to make kerfs about ¾ in. deep on the underside of the plate in line with the trimmer studs, as shown in the drawing. On long or heavy wall sections, I limit the kerf depth to about ½ in.

After the wall is assembled and erected, it's a simple matter to handsaw out the plate. This method saves time and the teeth on my favorite handsaw when I'm building on concrete floors, and still keeps the framing properly spaced and rigid as it's being raised.

—Brian P. Mitchell, Somerset, Colo.

Perpendicular toenailing

About 15 years ago, I started using a toenailing trick that makes it easy to start the nail without its slipping. I put the nailhead where I want to toenail, as shown in the drawing, and rap the point a couple of times with my hammer. This makes an angled flat where I can drive the nail at right angles to the wood surface. It also blunts the nail, which makes it less liable to split the wood.

—Andrew Hamilton, Antioch, Calif.

Hanger notch

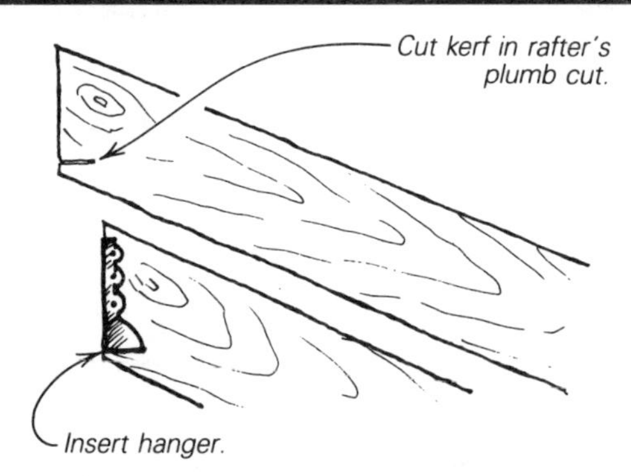

When I have to attach a rafter to a beam with a joist hanger, I find it helpful to kerf the rafter's plumb cut with a skillsaw. Then I can insert the hanger into the kerf, as shown in the drawing, and nail it in place. The kerf depth coincides with the full cutting depth of my 7¼-in. saw, so I don't have to mess with adjustments. I can also stack my rafters and make one cut to kerf them all at the same time. This method is a lot easier than notching the rafter, and it provides a continuous plane up the rafter to attach gypboard. —***Greg Halverson, Portland, Ore.***

Corner blocking

Until recently, framing walls with 45° angles had been something of a pain. I'd always end up with an odd-shaped void right on the corner on the outside face of the wall. Just leaving it would bring threats of bodily injury from the drywallers, so we'd have to cut an even more peculiar piece to fill this annoying gap. Then we discovered a way of filling the corner easily, with a minimum of waste.

First we rip a 4x4 diagonally on the bandsaw (this can also be done on the 10-in. table saw but it takes two passes). Then we set the table-saw fence at 3½ in. and rip each of the halves, long side down, as shown in the drawing. This leaves us with two pieces that have a kite-shaped cross section—two 1½-in. faces and two 3½-in. faces. This kite shape precisely fits that odd space at the corner.

These pieces are especially useful in bays, which need a

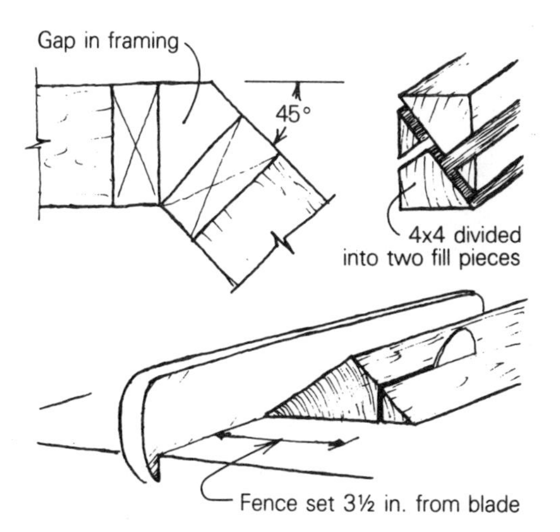

maximum amount of window space and a minimum amount of framing. Since the corner piece has virtually the same bearing area as a 2x4, we use them as king studs between trimmers. —***Tim Pelton, Fairfield, Iowa***

More corner blocking

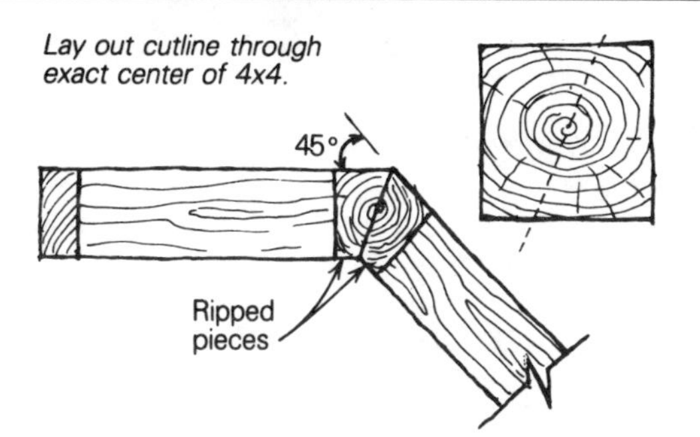

To block 45° corners, I rip a 4x4 on the bandsaw as shown in the drawing, with the table set at 22½°. Then I use the ripped pieces at the ends of each wall section. This trick wastes no wood, and also allows the sections to be securely nailed together. —***Ron Milner, Grass Valley, Calif.***

Raising walls with jacks

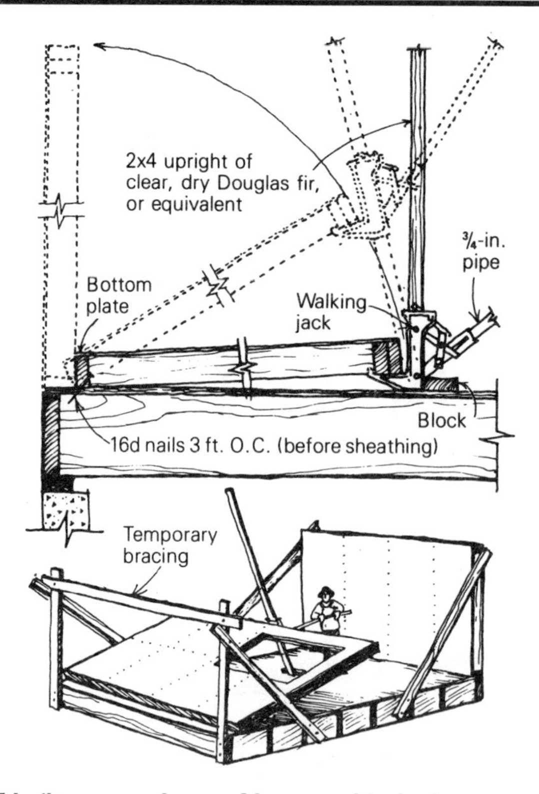

When I built my own house, I borrowed jacks from my lumberyard and raised the walls myself. My lumberyard doesn't lend the jacks anymore, but they sell them for $95. In Portland, they can be rented for about $8 a day. The jack has a slot that accepts a 2x4 upright and sharp clamps, called dogs, which grip the 2x4 to raise the jack when the lever is operated. A ¾-in. pipe inserted in the jack acts as the lever.

Build the wall section on the floor, but before attaching the exterior sheathing, toenail the bottom plate to the subfloor with 16d nails 3 ft. on center. The nails, bending as the wall is raised, will act as a hinge. Nail a block to the decking behind the upright, so the jack won't slide out of position. Temporary bracing keeps the wall section from continuing past vertical. I raised 20-ft. sections of wall with one jack and 30-ft. sections using two. The jacks are equally good for lifting heavy beams into place when used with rope pulleys and hoists. —***Walter Holmstrom, Lake Oswego, Ore.***

Solo framing

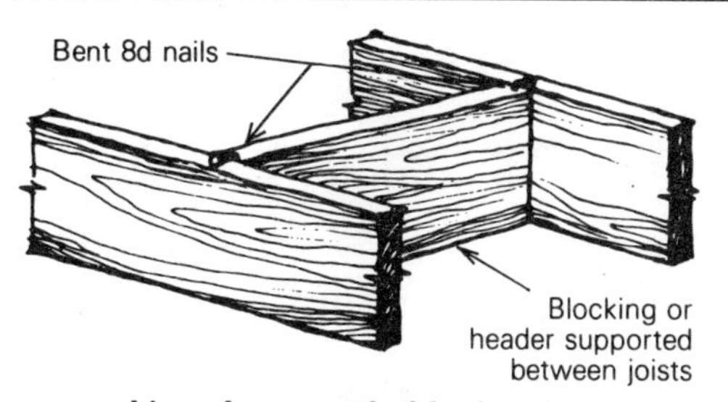

A carpenter working alone can hold a header, blocking or similar framing member in place for final nailing with a few bent nails. I use 8d nails, driven about a third of their length into the top edge of the work and then bent 90°. These ears will support the piece until the first nail is set.

—Craig Savage, Hope, Idaho

Wall jacks

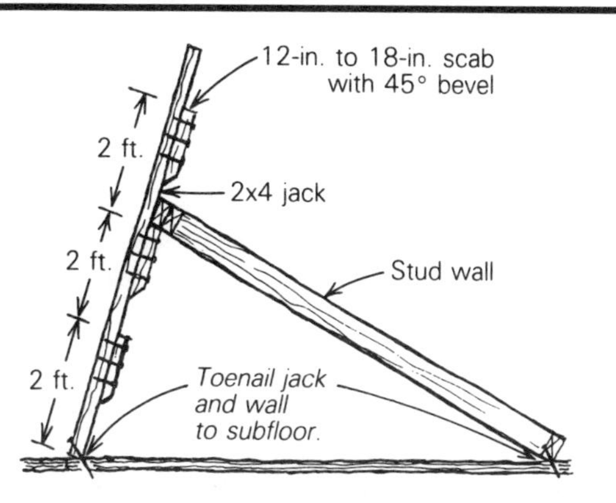

With three or four of the site-built wall jacks shown in the drawing above, a two-man crew can lift 30-ft. to 40-ft. long stud walls. Each jack is made of a 2x4 about 7 ft. long, with three nailed-on scabs, 2 ft. apart. The scabs form ledges to support the wall as it is lifted.

With the stud wall lying flat on the subfloor and toenailed to it, toenail each jack to the floor so that its base is tight to the wall's top plate. Now begin at one end of the wall and lift it to the first notch. Move back and forth along the wall, lifting one notch at each station until you've reached the third level. From there it's an easy push to get the wall upright.

—Ed Wilson, Seattle, Wash.

Scaffold jacks

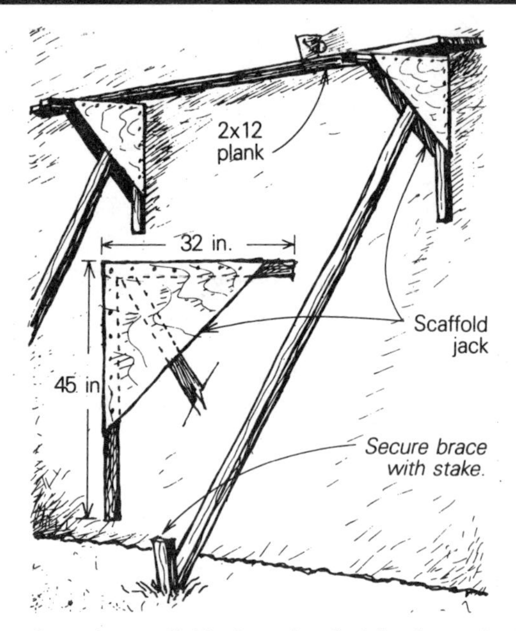

Whenever I need a scaffold of modest height, I use the jacks shown in the drawing above. They're easy to construct, don't cost much and they've never failed me.

The jacks are held in place by clear 2x4 braces that angle up from grade. Braces to the ground should be staked. If you're working off a subfloor, a nailed-down cleat will securely hold a brace.

The jacks are made from 2x4s and ½-in. plywood. The horizontal leg should be about 32 in. long, while the vertical one should be about 45 in. long. For added safety, I glue and nail the gussets to the legs. I use a 2x12 plank on top of the jacks, and I've used them in series with lots of jacks and planks to handle long runs.

—Dave Blom, Bellingham, Wash.

Double-beveled rafters

Carpenters in central California use this technique for cutting double bevels on hip or valley rafters fashioned from conventional 2x framing lumber. Set the tongue of the framing square on one end of the rafter as shown and draw two parallel pitch lines (they will be 1½ in. apart). Then

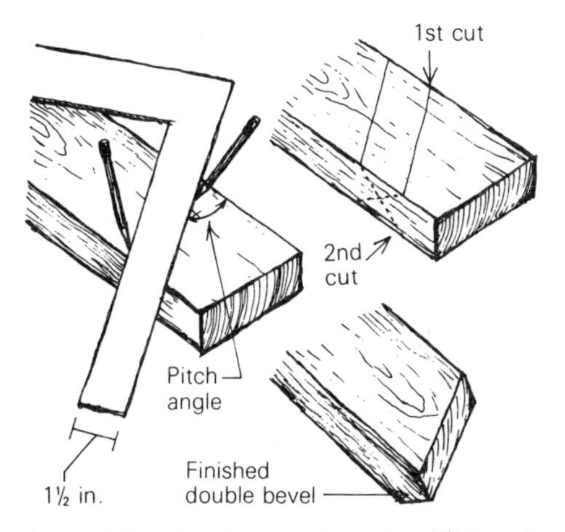

adjust the foot of the circular saw to cut a 45° bevel, and make the first cut in the direction indicated in the drawing. The second cut starts from the opposite side of the board. The resulting double bevel allows the carpenter to tuck the rafter between two perpendicular common rafters.

—Andrew Kujawa, Santa Cruz, Calif.

Double cuts

Here's a quick way to get a tight joint with ¾-in. thick rake boards where they meet at the roof peak. Tack your two rake boards in place on the gable end, letting them run wild at the top, one over the other, to form an X. Mark your plumb cut. Set the proper depth on your saw and your shoe angle at 7°, and cut through both boards at the same time. You want the bevel in the top board to be an undercut, as shown, so lap the boards accordingly. This will be a function of which way the shoe on your saw tilts, and whether you are cutting from above on the ridge or from below on a scaffold or ladder. This trick also works on straight cuts to join long runs of fascia. ***—Matthew Giordano, Plymouth, Mich.***

False beams from prefab trusses

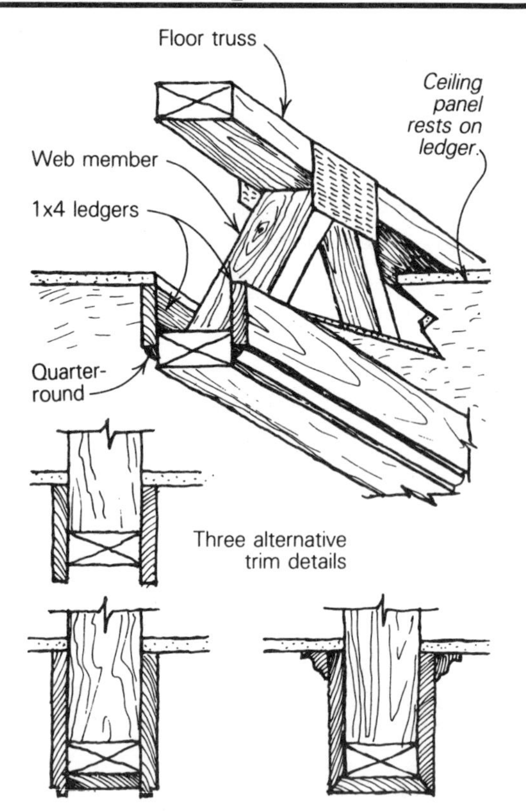

Floor trusses have several advantages over floor joists. They span greater distances, they are more uniform and easier to work with, and they eliminate the need for all the drilling it takes to accommodate pipes, wiring and ductwork. On the other hand, trusses typically take up more room than joists. In our house, we had to use 12-in. deep floor trusses to build our basement pool-room addition. If we'd gone with floor joists, we could have used 8-in. deep members. Since we wanted an exposed-beam look in the room, our problem was somehow to recapture the extra space required by the trusses, and to avoid grafting on phony hollow beams that would give us the effect we wanted but rob us of needed headroom. Here's how we did it.

We left the bottom chord of each truss exposed, and nailed knotty pine 1x4 ledgers to the web members, as shown above. This left a ¾-in. reveal, which we detailed with ½-in. quarter-round molding. We painted the trimmed-out trusses with a wash of mustard-color latex paint, which ties the colors of pine ledgers and trim and the fir trusses together, and lets the grain show a bit. Depending on the level of finish you want to achieve, the other trim details shown in the drawing give you a range of options.

The final step in this project was to lay a ceiling on top of the beams. We used random-width Armstrong ceiling panels, and simply laid them on top of the ledgers. No nails or staples were needed.

—Phyllis Brubaker Pyle, Fleetwood, Pa.

Energy-saving sole plates

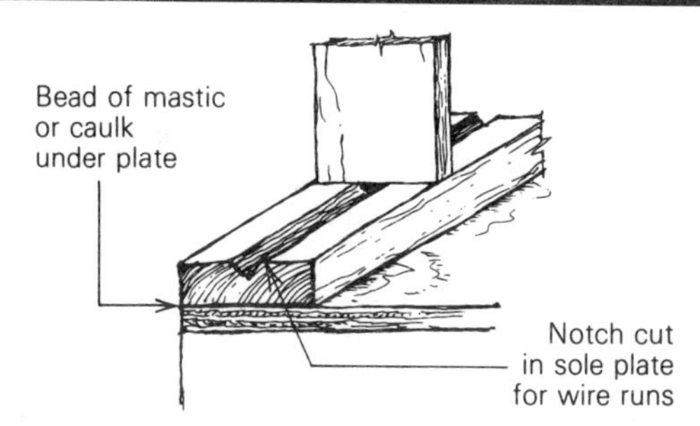

The usual method of wiring a stud-wall house is to drill holes in the center of the studs and string the wire around the perimeter of the building. When the insulation is installed later, the insulation adjacent to the wiring is compressed, lowering its effectiveness.

A more efficient and less cumbersome wiring scheme, which I learned from Larry Medinger of Ashland, Ore., is shown is the drawing above. A V-groove is cut in the sole plate before the wall is assemble. This groove makes an easy run for the wiring and is out of the way. A bead of mastic or caulk under the plate will cut down on air infiltration, a well-known energy bandit.

—David Bainbridge, Berkeley, Calif.

Chalkless nailing gauges

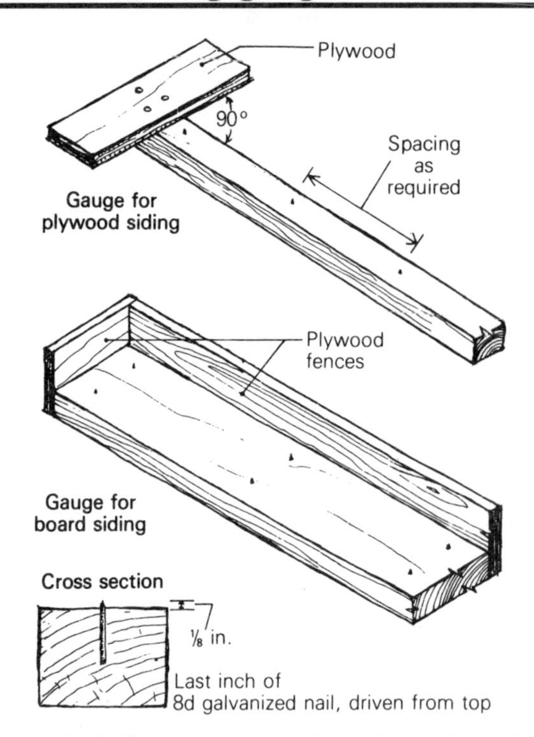

These two site-built gauges for marking plywood and board siding for nailing come in handy where nail lines and spacing have to be uniform, or where visible chalk lines cannot be left on the siding.

The nailing gauge for plywood is like a very long T-square. The blade of the square has nail points on its centerline at the specified nailing schedule. We found that the last inch of an 8d galvanized nail, driven blunt end first, worked best when the point was left to protrude about 1/8 in.

Allowing for the offset between the nail points and the edge of the gauge, lay out stud centers on the plywood top and bottom, hook the marker over the top edge of the plywood (plywood should be face up) and along the layout marks, and strike the device with a hammer along its length. The perforations that will be left represent the nail holes to be used in attaching the siding.

The gauge for board siding is made from 2x framing lumber (as wide as the boards to be marked) with plywood fences nailed to one edge and to one end. The nail points in the 2x are spaced on stud centers along its length, and as desired across its width.

The siding is marked by positioning a precut board against both fences face down and applying pressure; once again, the perforations indicate nail locations. Boards can be positioned and marked at various points along the gauge to allow for joints in the siding or openings such as windows and doors. Nail temporary stops on the gauge to locate these positions. These easily made jigs enable you to produce accurate, evenly spaced nailing lines at production speed.

—Malcolm McDaniel, Berkeley, Calif. and Paul Spring, Oakland, Calif.

Scissor framing

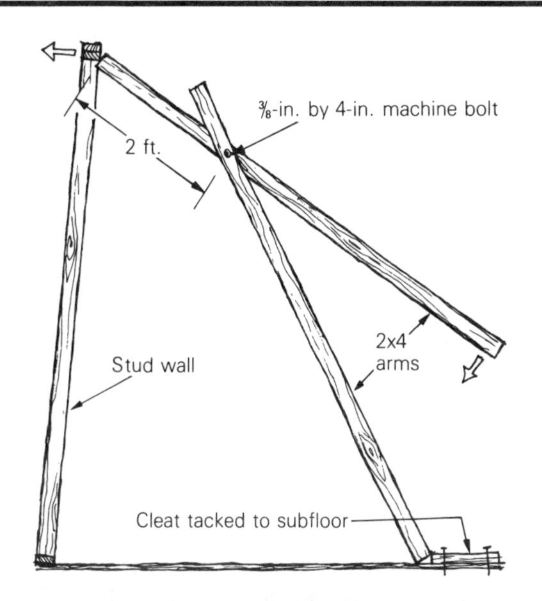

The scissor-like tool illustrated above can be useful in plumbing and aligning stud walls during framing. With it, a worker can exert and maintain great pressure against a wall while another worker nails bracing to hold the wall in its correct position.

The scissor can be assembled quickly on the job site from a pair of 8-ft. 2x4s and a ⅜-in. by 4-in. machine bolt. The proportions shown in the illustration are only approximate; I usually trim off one end or the other until the tool feels right and gives the best leverage.

To use the scissor for plumbing a wall, tack a cleat to the

subfloor about 6 ft. from the bottom plate and place the foot of the tool against it. Next, open the arms of the scissor until its raised end can be firmly wedged against the top plate of the stud wall. By pressing down on the tool's handle, you will increase its span and force the top of the wall away from you. Fine adjustments can be made by altering the pressure on the handle. ***—Malcolm McDaniel, Berkeley, Calif.***

Save your siding

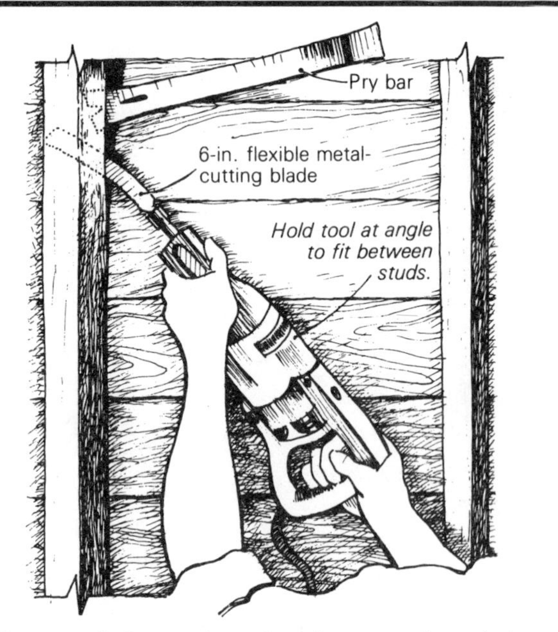

I recently expanded a portion of my house and needed to remove an existing exterior wall. I wanted to match the original siding, so I searched local lumberyards for the same pattern. I eventually found the proper sheathing and was staggered by the price. I realized I was about to remove close to $300 worth of perfectly good original siding from my house. The problem was how to baby it away from the old studs without rendering it useless. I tried prying, without success; the brittle old boards cracked and splintered with such regularity that it became obvious that another method was essential.

The solution was to cut the nails from the inside with a reciprocal saw mounting a metal-cutting blade. I have Milwaukee Electric Tool Company's Sawzall (13135 W.

Lisbon Rd., Brookfield, WI 53005), but any make will do. Remove the shoe from the tool so you can angle the blade nearly parallel to the siding. I used 6-in. blades, bent slightly to allow for the difficult angle. I found a little wedging with a pry bar made it easier to insert the blade between the siding and the studs. Be careful to keep the shaft of the saw away from the siding or studs. Contact can result in costly damage to the saw's bearings, gears and the shaft itself.

Be sure to wear safety goggles, because the blade will occasionally snap during the work. I must have broken $10 worth, but I sure didn't mind when I had my $300 worth of siding in usable condition. I figure it paid for the Sawzall, as well. ***—Hadley Green, Santa Monica, Calif.***

[Editor's note: Reciprocal saw manufacturers offer an offset blade adaptor that allows close-quarter sawing without removing the protective shoe; Milwaukee Electric Tool Company offers attachment #48-03-2000.]

Wall bracing

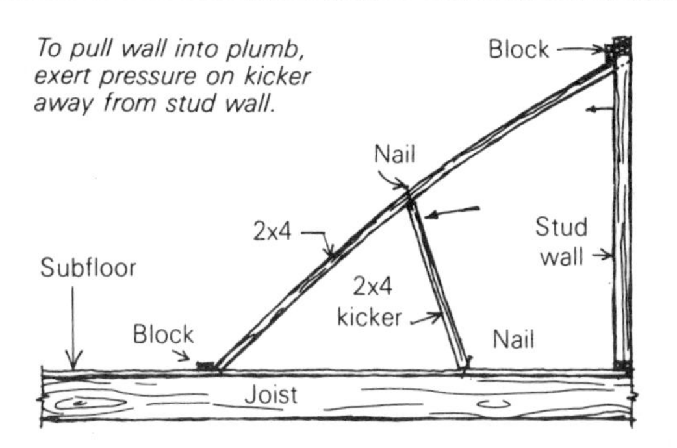

One common problem in framing is how to straight-line a second-story stud wall and brace it for joisting or rafters if it leans out. Here's one way you can pull the wall back into line and keep it there.

Lean a 2x4 (face up) against the stud wall with one end resting on the top plate and the other end on the subfloor at the location of a joist. Nail both ends securely. For extra security nail a 2x4 scrap block flat against the double plate at the top of the brace to prevent it from coming loose under pressure.

Next, toenail a 3-ft. to 4-ft. 2x4 to the floor along the same joist midway between the wall and the foot of the brace. You'll get the desired results by pushing this kicker against the brace and exerting whatever pressure is required to draw the wall into line. Secure it by driving two nails through the brace into the end grain of the kicker. These braces can be left in place until the rafters or joists are installed.

—Marc Davis, Tucson, Ariz.

Splined siding

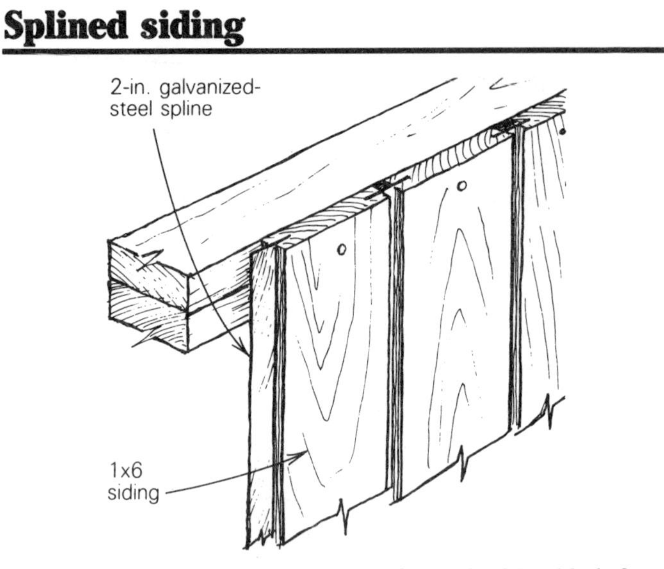

I recently built a new workshop and I sheathed it with 1x6 boards joined with galvanized splines. I like this method because it gets the most out of the width of a board, and the splines can often be scrounged for next to nothing. I used 2-in. wide galvanized steel that is commonly used here in Australia for strapping together large shipments of bulky goods. Thin strips of nylon or plastic could also be used.

My wall framing consists of 4x4s on 8-ft. centers, with plates at top and bottom and a mid-rail. I ran each piece of siding over my table saw to cut a 1-in. deep kerf in each edge. Once I had a board in place, I inserted a full-length spline into the kerf, as shown above. The neighboring board accepts the protruding spline, and so forth. I kept the nails away from the splines, and left a little play between adjacent boards to allow for expansion of the wood.

—Jonathan Davies, Queensland, Australia

Trimming siding

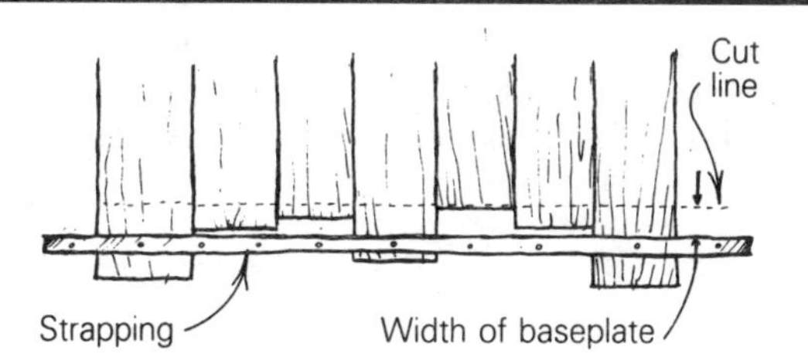

Whenever I finish a building's exterior with vertical tongue-and-groove, shiplap or rough board siding, I use the method shown above to make straight, neat cuts. Every 3 ft. or 4 ft., I leave a board long by 4 in. to 6 in. Then I snap a chalkline about 1¾ in. down from my intended line of cut. This dimension is the width of my saw's baseplate from its edge to the blade. A piece of strapping tacked below this chalkline gives me a ready-made saw guide that will produce a crisp, straight line on the board ends.

—Bruce MacDougall, Bridgewater, N. H.

Clapboard gauge

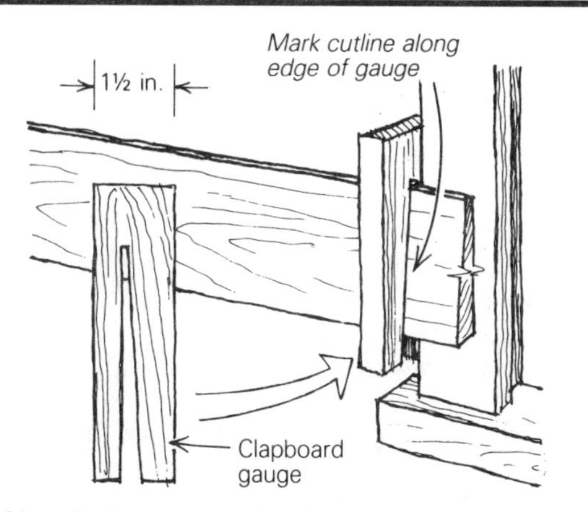

The marking device shown in the drawing above is used to establish cutlines for clapboards where they abut trim pieces. I find it to be especially handy for recladding old houses where the casings and cornerboards aren't plumb.

I make the gauge from a scrap of ¾-in. thick pine, 1½ in. wide. The slot in the center of the gauge should fit the profile of your clapboards just tightly enough to hold a clapboard in place.

To use the gauge, hold your clapboard with the end to be cut running by the casing, and slip the gauge over the clapboard. Press the gauge tightly against the casing, and scribe along the edge of the gauge. Cut along the line, and the clapboard will fit tightly against the trim board.

—Duke York, Willimantic, Conn.

Basement baseplate

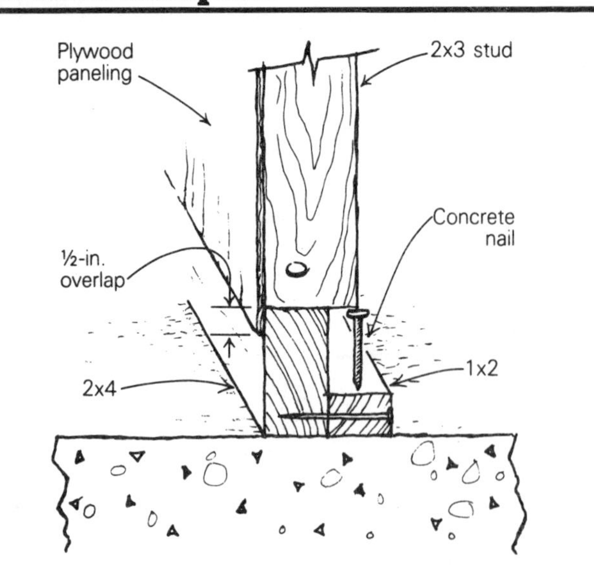

Here's a simple and economical way to create an attractive baseboard for a one-sided interior wall in a basement by using a composite bottom plate. It consists of a 1x2 nailing strip attached to a 2x4 placed on edge, as shown in the drawing above.

First, drill the 1x2 strip at 12-in. intervals with a bit slightly smaller than the nails you plan to use, and fasten the strip to the 2x4. Then paint the face of the 2x4 with flat black paint, and fasten the assembly to the slab floor with concrete nails. Now toenail 2x3 studs to the base on 16-in. centers, and panel the wall. Let the plywood run over the base plate about ½ in., leaving a flat black reveal 3 in. wide.

—John Molnar, Moorestown, N. J.

Useful prying

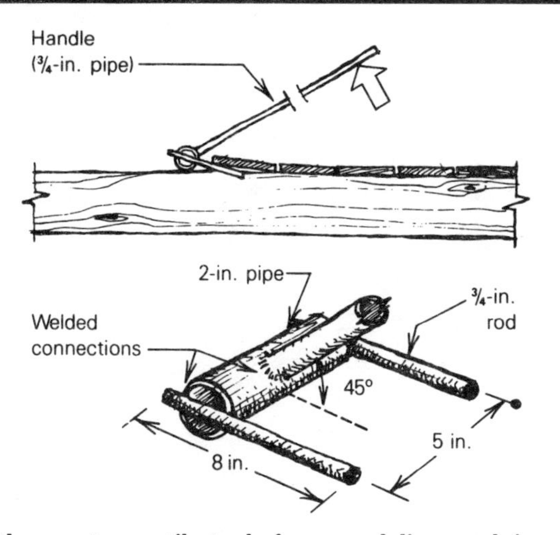

One of the most versatile tools for remodeling work is an ordinary, garden-variety mattock. The wide, flat blade is perfect for separating studs, rafters, joists, or anything else that is nailed together. In addition, it can be used to take up oak or pine flooring without destroying the tongues and grooves, and it can greatly speed the removal of shingles, roofing boards, plywood, bricks and a variety of other materials. Make sure you get a mattock with a blade that is fairly perpendicular to the handle; this makes it much easier to use.

Another very worthwhile tool is a homemade rig built from two pieces of pipe and two pieces of steel rod, as shown. This tool is excellent for removing roofing and siding boards without cracking them. Simply straddle the rafter, stud or joist and lift. The boards just pop up.

—Norman Rabek, Burnsville, N. C.

Chapter 9:

Floors

Plywood persuader

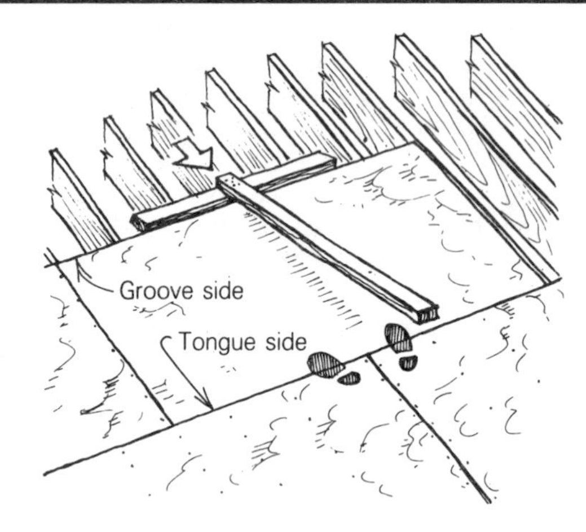

Laying tongue-and-groove plywood subfloor usually calls for at least two carpenters. As one wields a sledgehammer on the sheet to be threaded, the other is easing the tongue of this sheet into the groove of the sheets in the preceding course by shifting weight from foot to foot.

The large T-square shown above can eliminate one of these jobs. It uses a 2x6 about 4 ft. long as a crossbar, and a

2x4 handle about 5 ft. long. With the plywood panel in position, run the crossbar out on the joist tops with the handle held only 12 in. off the deck. Then pull it back with a lot of force against the grooved side of the panel while keeping the balls of your feet on the seam to be threaded.

For real efficiency, try a crew of three. One spreads the glue on the joists and flops down the plywood, the second person threads the sheets into final position with the T and tacks down the corners, and the third nails them off with a pneumatic nailer before the glue dries.

—Malcolm McDaniel, Berkeley, Calif.

Salvaging floorboards

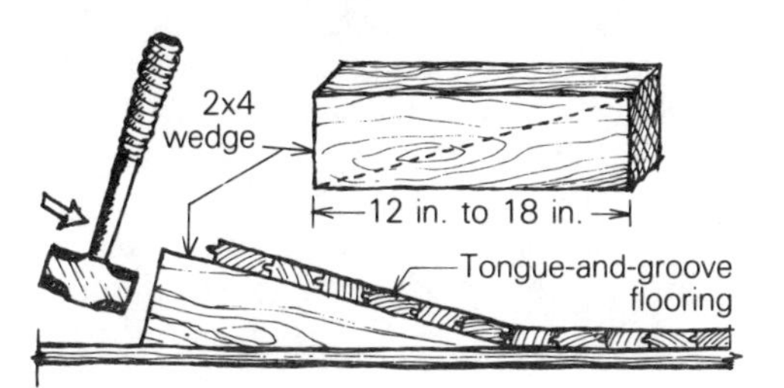

Trying to rescue a tongue-and-groove floor without ruining it can be tedious and time-consuming. The technique that I've developed keeps damage to a minimum, saves a lot of time and prevents frazzled nerves.

First, cut 12-in. to 18-in. 2x4s into wedges, as shown in the drawing. All wedges should be of equal length; the wider the floorboards are, the longer the wedges should be. Flooring has to be removed in the reverse order in which it was laid. Remove a few runs of flooring with a prybar to expose the subfloor, and position the wedges about 12 in. apart, cut side down. Beginning at one end, drive the first wedge under the floorboards about an inch. It's essential that the wedges be advanced slowly, or the flooring will split. Move the next wedge and repeat the process. This way the flooring is gradually lifted onto the wedges and pulled free from the subfloor.

When the loose flooring reaches the driving end of the wedges, remove a few strips and lay them aside for future cleanup. If a wedge hangs up on a nail, simply reposition the wedge.

—A. Eugene Walbridge, Easton, Md.

Leaf-spring nailset

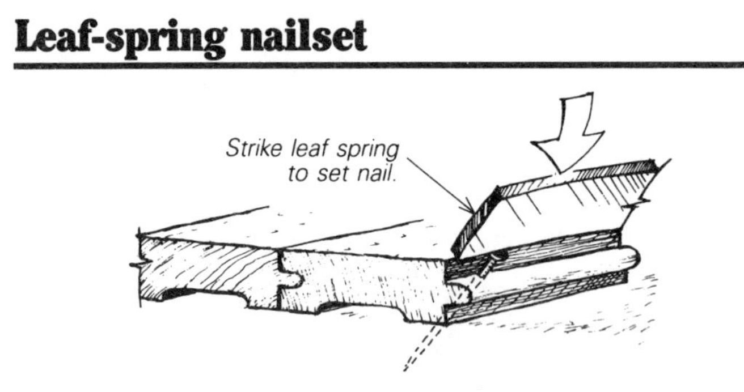

If you don't get a lot of practice laying tongue-and-groove oak or Douglas fir flooring, chances are you'll need a nailset to help drive the nails to their proper depth. Nailsets are expensive, and dull ones tend to slip. Using one for a big job like a floor will shorten its life considerably.

My alternative is a piece of upper leaf spring from an old car, used as shown in the drawing above. It's hard enough to take the blows without deforming, and it's large enough for you to get a good grip on the tool and still keep your fingers out of the way. The slight curve of the spring keeps most of the edge away from the flooring while it is being used.

—Carlyle Lynch, Broadway, Va.

Squeaky subfloor fix

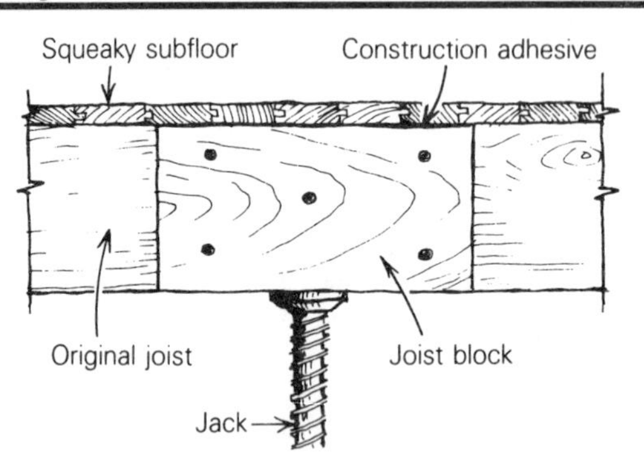

To quiet squeaks in floors, I use a small joist block sistered to the original joist, as shown in the drawing above. Once I've located the squeak, I put a good bead of construction adhesive along the top edge of a block that is the same depth as the existing joist. With the block in place, I use a small jack in the crawl space to press it tightly against the subfloor. Nailed off and jack removed, the job is done.

—Ross Fulmer, Atascadero, Calif.

Reluctant decking

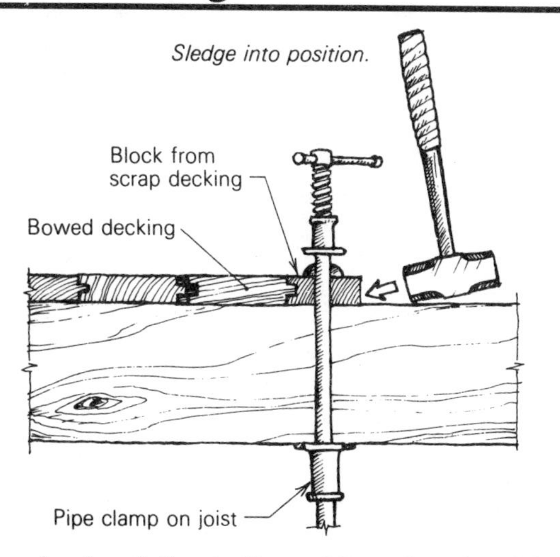

Here is a simple solution to the problem of getting tight joints when laying twisted or bowed tongue-and-groove decking.

First, pick some good straight pieces for your first course or two and nail them securely in place. Working with the tongue forward, start the groove of the next course onto the tongue of the last course. Anywhere the pieces don't slip together easily or a bow in the wood causes a space to occur, clamp a block (a short piece of scrap decking works well, minimizing damage to the tongue) to the joist in front of the troublesome piece with a short pipe clamp. Then hammer in on the block with a sledge hammer.

The tougher the blow, the harder you will have to clamp the block down and the harder you will have to hit it with the sledge, but the clamped block will hold the joint tight until it can be nailed. In this manner you can lay the most twisted pieces easily and tightly.

This trick was shown to me by Joe Caputo, an old-timer I work with building post-and-beam structures. We use it quite often to apply 2x8 V-match pine to 6x8 floor joists 3 ft. on center. We use the V-match as either a subfloor or the finish floor, and also as the finish ceiling for the room below. This technique works great on the timbers and equally well on stick-framed buildings. For fastest application try a toenail first, and if that won't pull in the decking, try the clamp method. ***—David Barker, Gardiner, Maine***

Double decking

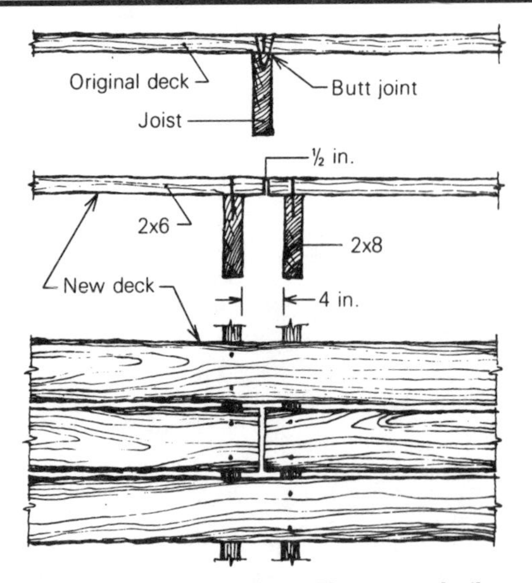

Our house has decks on two sides. They were built conventionally: 2x8 joists set on 4-ft. centers covered with 2x6 planks set on 6-in. centers. When I had to replace the decking, primarily because of rot that started wherever two 2x6s butted each other over a joist, my design involved doubling up the joists at each station, as shown in the drawing. This simple change accomplishes the following: no tight butts in planking, no butts over joists, no split 2x6s from nailing them at their ends, and the entire structure can now "breathe."

The deck has a width of 15 2x6s. With careful layout using lengths up to 20 ft., I have no more than four joints at any station, minimizing waste. I used pressure-treated southern yellow pine. All the lumber cost about $1,000, and only $64 was added by doubling up on the 2x8s.

—James B. French, Portsmouth, R. I.

Persuading plank flooring

When carpenters install plank flooring, they often have to cope with warped planks. Some use a prybar to bend reluctant boards, others use pipe clamps. I keep it simple and use a wedge made from the decking itself.

As shown in the drawing, I make a diagonal cut through a scrap piece of decking that is about 1 ft. longer than the on-

center spacing of the joists. The tapered piece with the tongue on it becomes the block, the other the wedge. Next, I nail the block to a joist just in front of the warped board. Then I drive the wedge, grooved side against the tongue of the bowed decking, into the space between the block and the floorboards.

When the warped floorboard is wedged snug against its neighbor, I blind-nail it in place. To allow blind nailing next to the wedge, I sometimes have to remove the top part of the groove from the wedge. ***—Stephen Reddy, Monterey, Mass.***

Deck jack

Instead of using pry bars or bump sticks to achieve a tight fit with T&G decking, we use a big bumper jack. Its base bears against a 2x4 block nailed to the floor framing, while the jack leans on a scrap piece of the T&G decking. Using this method we get a tight fit every time, with T&G boards or plywood, even in the winter when the decking stiffens up in the cold weather. ***—Joe Eller, Freeville, N. Y.***

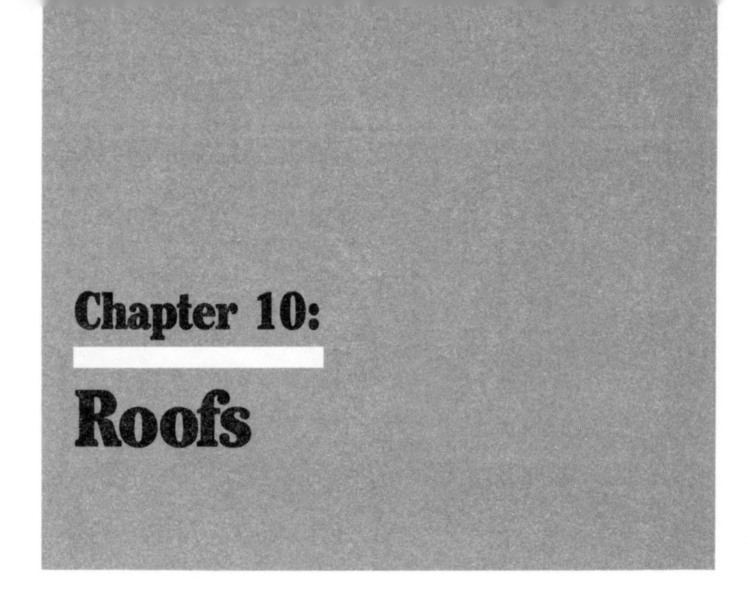

Chapter 10: Roofs

Roof-sheathing jig

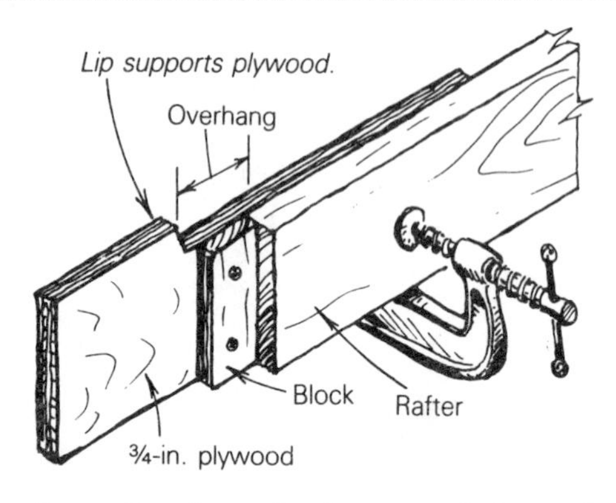

Whenever I'm nailing down the first row of plywood on a roof that will have soffits, I use a pair of jigs shown in the drawing to position the panels. Using these jigs, I can adjust the amount of overhang to suit the fascia detail by moving the blocks in relation to the lip that supports the edge of the plywood. I've found them to be especially helpful when I am working alone and in need of a third hand.

—John Shepherd, Charlottesville, Va.

Ridge-vent jig

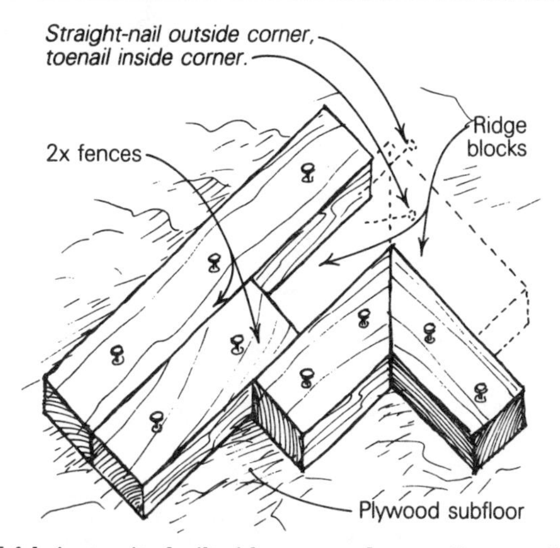

When I fabricate site-built ridge vents, I use a jig to make quick and accurate 2x ridge-block assemblies. Once I know the pitch of the roof, I set my miter saw accordingly and cut the blocks for either side of the ridge. Then I face-nail a pair of these blocks all mated up to a firm flat surface, such as a plywood subfloor. This forms the bottom of the jig. Just above one leg of the pair, but well back, I nail two pieces of 2x scrap as shown in the drawing. These act as fences to help align the blocks and to hold them steady as I nail them together in the jig. A 16d nail driven straight through one block into the other on the outside corner and a toenail near the inside corner are all it takes to join the two blocks, and they're ready for installation on the roof.

—Jim Halls, Johnson, Vt.

Shingle removal

Most of the older houses where I live have asphalt shingles laid right on top of the original wooden shingles. When it's time to tear them all off in preparation for a new roof, most roofers use a flat shovel to pry off all of the old layers of shingles. This is hard, slow work because the shovel keeps ramming into the thousands of nails securing the shingles.

Instead of a shovel, I use a short-handled, four-pronged spading fork with heavy, flat tines. The tines slide around the nails instead of ramming into them. After I remove the old ridge caps, I start at the top of the roof and work across it horizontally. The constant levering action puts a fair amount of strain on the tines, so I check them occasionally to make sure they stay parallel with one another. If one gets bent out of alignment, shingles tend to get wedged between the tines. To bend the tines back in line, I use a length of ¾-in. water pipe. —***Steve Funcell, Jamestown, N. Y.***

Shingle cleanup

Stripping the old shingles off a house is a messy job by anybody's standards. The accumulated pieces of cedar or asphalt tend to rain down around the house, often damaging fragile landscaping. To prevent this problem, I use a sheet of 6-mil poly to act as a combination protective barrier and tarpaulin, as shown in the drawing above. I staple the poly to the top piece of a 2x4 frame, making sure that the top edge of the poly is doubled and wrapped around the horizontal 2x4. This arrangement saves a lot of cleanup time, and leaves my clients smiling. —***Charlie Woodhouse, Kensington, Calif.***

Asphalt-shingle re-roofing

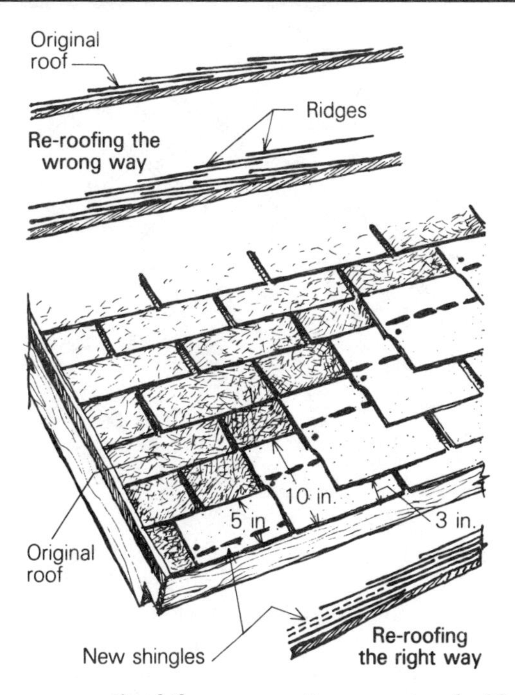

Have you ever noticed the wavy pattern on a roof with multiple layers of asphalt or fiberglass shingles? This is due to the uneven thickness of the shingles where they overlap, and it detracts from the appearance and life of the roofing. On a new roof, the overlap doesn't cause a problem, but when a second roof is applied in the same manner as the first, the overlap creates as many as six thicknesses. This causes ridges that get larger as each new roof is applied, as shown in the drawing above.

To prevent shingle buildup on the typical 5-in. exposure roof, trim the first row of new shingles to 5 in. and the second row to 10 in., and butt them against the bottom edge of the old courses above. These will work as spacers that allow the new roof to start a different overlap pattern that misses the old bumps, with a 3-in. exposure on the first row and a normal 5-in. exposure from there on. This method lets you butt the new shingles against the bottoms of the old. But first make sure the courses on the old roof are straight.

Your new roof will now lie flat, look great and last longer because the wind and water won't have any bumpy ridges to erode. —***Jack McGhie, Tucson, Ariz.***

Buttoned-up venting

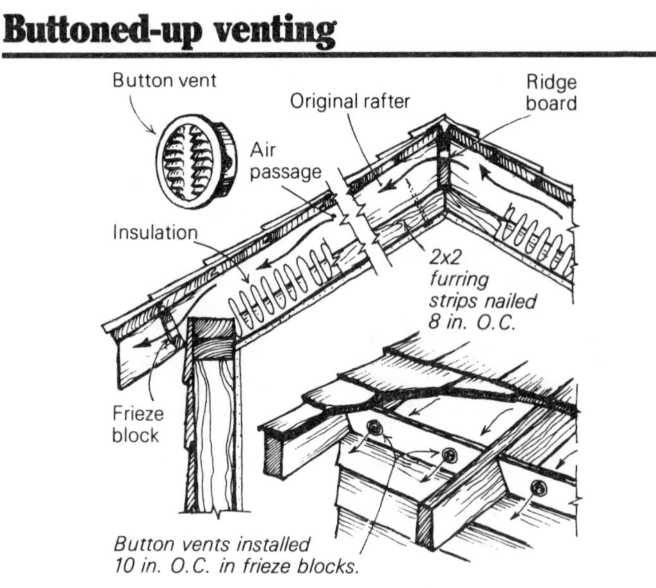

Button vents installed 10 in. O.C. in frieze blocks.

I remodeled our kitchen last year and opened up the previously flat ceiling to reveal the pitch of the 2x4 rafters. In its original state, the attic had been without insulation and vented by generous screened openings at each gable end. I wanted to insulate the ceiling, yet still provide adequate space for air movement to minimize moisture accumulation around the insulation.

My solution was to nail furring strips of scrap 2x2 lumber along the bottoms of the rafters. Stapling 3½-in. thick fiberglass batts to the furring strips left a 2-in. airspace above the insulation. But where would the airflow come from? On our house, there are frieze blocks between rafters above the outer wall. They act as both blocking and exterior finish. I simply installed 1½-in. button vents 10 in. on center in the frieze blocks. Button vents (or plug vents) are round, aluminum fixtures with tiny louvers that keep out rain and insects. They are easy to install and available in a number of diameters. I used a hole saw to get through the 1x4 frieze blocks, because 1½-in. drills tended to split the old dry wood. To assure uninhibited air flow between opposing rafter bays, I drilled two ¾-in. holes in the ridge board, 10 in. on center. —***Alex Westrom, Wilmington, N. C.***

Button vents revisited

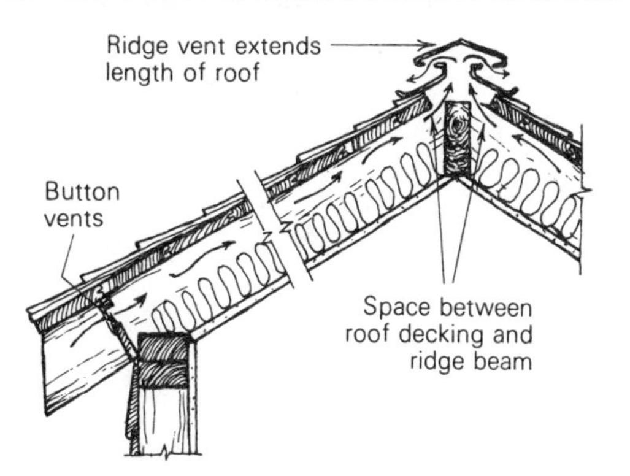

The previous method for venting the space between insulation and the roof of a cathedral ceiling is based on using button vents on both sides of the roof at the bottom of the rafters. Supposedly this would allow a flow of air in one side, up to the peak, down the other side and out. This is not an effective method of ventilation. Hot air will gather at the peak and stay there unless ridge venting is installed to let it out, as shown in the drawing; cool air will then replace the vented warmer air, allowing the desired air flow. Tables are available with specific vent dimensions for the area to be vented. Local building departments can usually supply them. ***—Steve Minkwitz, Marblehead, Mass.***

Holding down a temporary roof

This tip is for the homebuilder who has run out of time, money or both and needs to get a roof covered before the next rain. I've tried it and it works on roofs with a minimum 4/12 pitch in any weather condition short of a gale.

After whatever sheathing you are using is in place, cover it with 30-lb. felt using any appropriate lap, then staple string ½ in. above the lap line. Use kite string, mason's line, fishing line, or some other small-diameter string with at least 15 lb. test. The staples should be at least ½ in. long (⅝ in. is even better) and place them 10 in. on center for effective holding. Be careful to staple every exposed edge, vertical or horizontal, or a good wind will pick off the felt. The string can be left in place when the finished roof is installed.

—Craig Savage, Hope, Idaho

Ridge-hung ladder

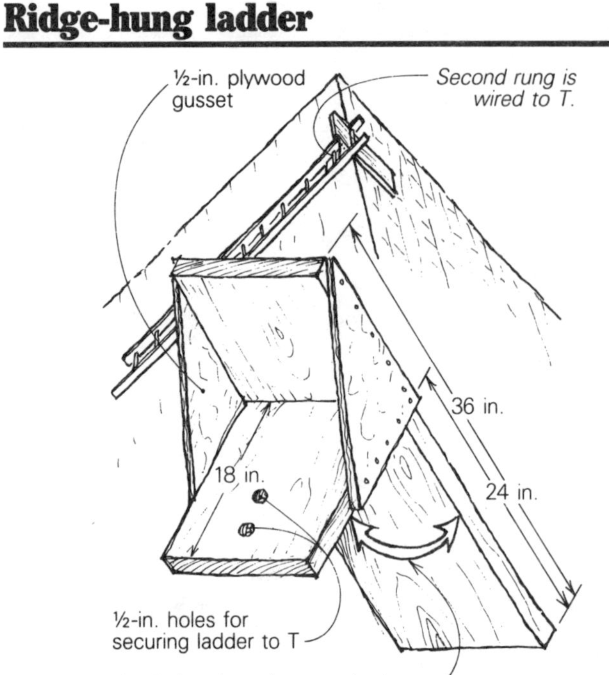

When we have to work on a steep pitch where nailed-on footholds are impractical, such as metal or fiberglass roofs, we like to stand on a ladder that hangs from the ridge. The top rung of the ladder slips over a wooden T-shaped piece, as shown in the drawing above. The T is made of two pieces of 2x12 joined with 16d nails and braced with ½-in. plywood gussets. The angle of the T should be adjusted to match the slope of the roof.

When we use this rig, we wire the second rung of the ladder to the T. This helps to keep everything parallel to the roof slope. We've found this arrangement to be very stable, and the ladder and its T can easily be moved along the ridge. For a no-scratch model for metal roofs, line the inside of the T and the rails of the ladder with glued-on carpet.

—Jim Greenfield, Golden, Colo.

Roof-gun tune-up

I use a pneumatic staple gun to install composition shingles, and every now and then the tool will clog up because roofing tar gets into the tip and the safety mechanism. Now I carry a can of Gum Out carburetor cleaner on the roof with me. Whenever the gun starts to get sticky, I give it a shot of the carburetor cleaner. It dissolves the tar without my having to disassemble the tool. Gum Out is available at auto-supply stores. *—Jeffrey A. Oswalt, Kimmell, Ind.*

Cutting sheet steel

I live in a small town in southern Colorado. Less than 20 miles from my house the average snowfall is a whopping 500 in., so metal roofs are very popular around here. The snow will slide off them before it gets a chance to pile up. Most of the roofs I install are made of sheet steel, with a baked-enamel finish. I have tried numerous ways of cutting this material, but the best way I have found is to use an old circular sawblade mounted backwards on the saw's arbor. This arrangement makes a fast, accurate cut. Be sure to wear safety goggles and ear plugs. This is a very noisy operation. *—Grover Hathorn, South Fork, Colo.*

Backing up a cut

Recently I had to make a hole in a lath-and-plaster ceiling for an attic fan. Of course the location of the opening was determined by what looked good from below, rather than by the joist spacing. The existing ceiling was in good shape, and I wanted to keep it that way. Needless to say, cutting the hole involved sawing through a large swath of plaster—well away from the joists and without their solid backing. I'd rather spend a little effort in the beginning of a project to reduce a lot of time-consuming repairs later on in the job, so I tried this improvisation.

First I marked the placement of the fan hole on the

ceiling, and then I drilled four small holes at the corners. This let me clearly see from the attic what was going to need support for the upcoming cut.

Next I mixed up a batch of molding plaster (plaster of Paris) and poured two lines of the mix along the crosscut path that the sawblade would have to take to get through the lath and plaster. While the plaster was still wet, I laid 1x4s flat on top of the mix, as shown in the sketch on the previous page.

Since the hole would need some picture-molding trim on the face of the plaster below, I went ahead before I cut the hole and installed this trim with screws that went through the plaster into the 1x4s. This captured the plaster in a clamping action between the trim and the 1x4s. Then I cut the plaster, lath and joists with a reciprocating saw.

—James E. Power, Lawrence, Kans.

Stepping up

I recently added three dormers on a 12:12 roof, using a toe-board and roof-stair system that saved me a lot of time, risk and effort. As shown in the drawing above, the stairs are made of two 2x12 stringers with 16-in. long, ¾-in. plywood treads. The uphill end of each stringer was cut in a curved pattern to keep the front edge from damaging the shingles when it slid onto the roof. The downhill end should be notched on the underside for a flat 2x4 toe board. I placed one ladder on each side of the dormer layout, and hooked them over the 2x4 toe board that was nailed at the eave parallel to the ridge. A 2x12 plank could then be laid across the treads of the two stairs, and moved up and down the pitch of the roof. Framing, siding, trimming, painting and shingling the dormers and surrounding area were relatively easy, and the system caused a minimum of damage to the existing roof.

—Wayde Millany, Dillon, Colo.

Shingle shelf

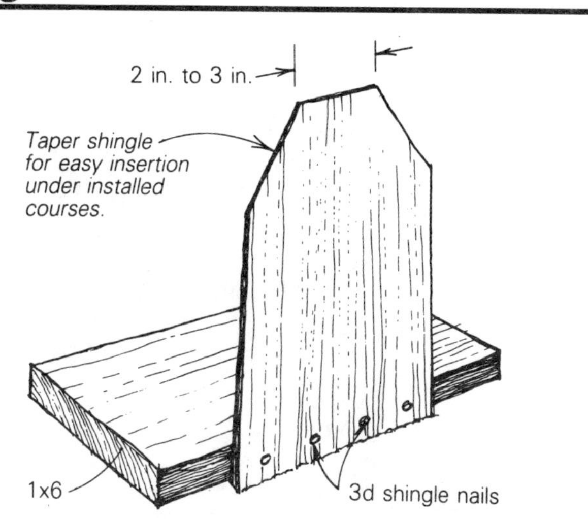

When wood shingling a roof, keeping a ready supply of shingles close at hand can be a problem. The simple shelf shown in the drawing above can be secured to the roof by tucking the tapered tab under an already nailed course of shingles. In this way shingles can be kept conveniently close to the height at which you are working, instead of down by your feet on the staging. A bunch of these shelves can be made from rejected shingles and scraps of 1x6s or 1x8s. Cutting the top corners off the shingle part of the shelf makes it easy to slip it up under a course.

—Kendall Gifford, Putney, Vt.

Valley flashing

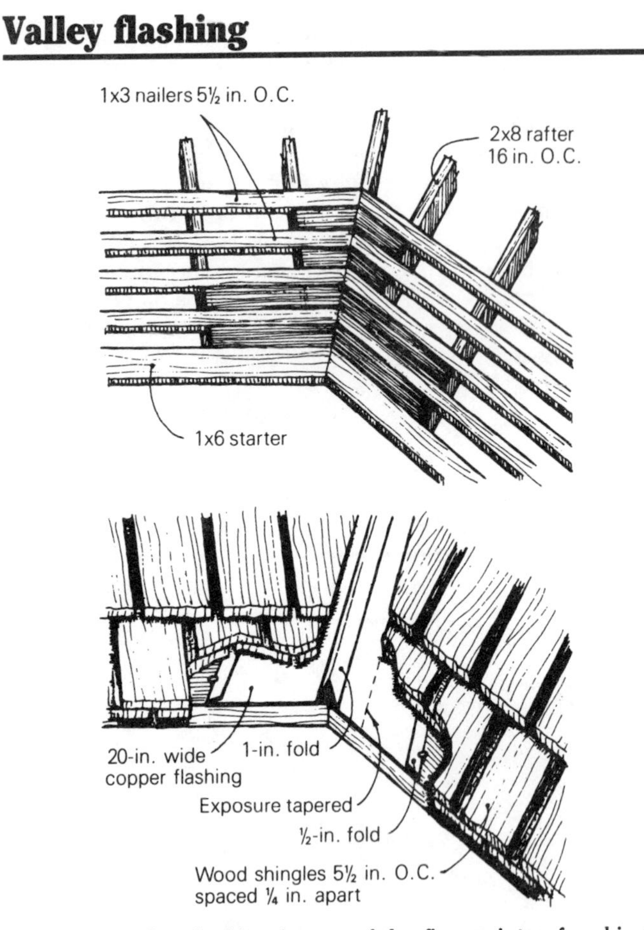

Successful valley flashing is one of the fine points of making a wood-shingle roof weathertight. I favor the old-fashioned method of nailing shingles into 1x3 horizontal stripping; fixing shingles to a solid sheathing such as plywood (even with a layer of felt paper between) doesn't allow air to circulate around shingles. Because most of the shingles I use have 5½ in. to the weather, my 1x3s are also spaced 5½ in., except in the valleys, where I butt strips together to create a solid base for the flashing that follows.

I use 20-in. wide copper roll flashing, which I cut to 4-ft. lengths on a shear. I bend the copper on a brake as shown. The middle folds first and then the outside crimps. The 1-in. fold in the center acts as a kind of levee to keep water driven in a storm from sloshing up under shingles on the opposite side; the ½-in. crimps are back-up protection in case some water does make it up underneath—the water can't jump the crimps so it runs back down the valley toward the gutter.

Placing and securing the flashing is another fine point. I overlap 4-ft. sheets about 6 in. to 8 in. for weathertightness. I fasten the sheets by snugging the heads of large-headed roofing nails tight enough against the outside of the crimp to pinch the fold. This keeps the valley sound and waterproof. I put a nail at every point that the copper passes over a rafter. Be sure when putting the lowest course of shingles in the valley not to destroy the copper's integrity by punching through it with a shingle nail. Careful work forms a very functional and lasting valley.

When placing the first course of shingles around the valley, I vary the amount of exposed copper from 3½ in. at the upper end to 5 in. at the lower end, a 1½-in. taper from top to bottom on each side of the center fold. This wider exposure at the bottom allows ice and leaves to slip out much more easily.

—***William C. Barthelmess, Woodbury, Conn.***

Extruded flashing

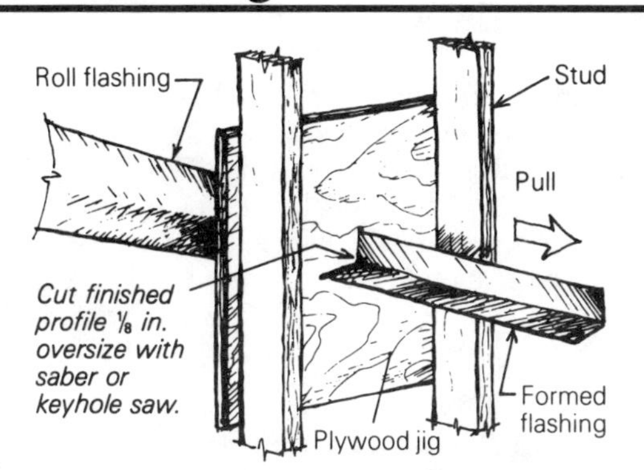

Bending your own flashing on site usually produces wavy, crinkled pieces of sheet metal that make you wish you'd had them made up at a shop. Here's a fixture you can make on the site that produces near-perfect results.

Using a jigsaw or a keyhole saw, cut the full-size cross section of the flashing you want to make into a plywood scrap that's at least 18 in. across. Nail the plywood across the studs of an unsheathed partition at chest level. Then cut a piece of roll flashing to length and insert it into one side of the cut in the plywood. With one person pulling and another supporting and feeding the flashing, this fixture makes the job easy. —***Mick Cappelletti, Newcastle, Maine***

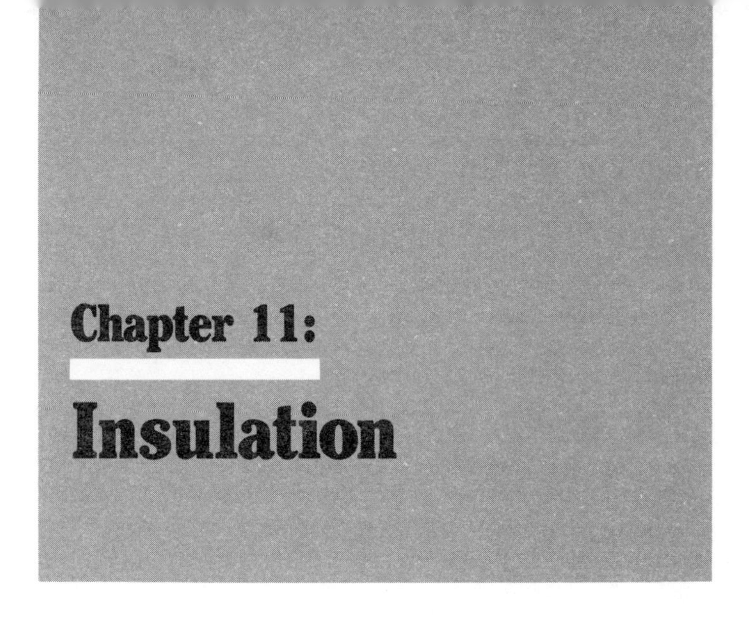

Cutting insulation

I recently insulated my entire house with foil-faced fiberglass batts, and I found a good tool for cutting the stuff down to size—hedge shears. That's right, but they have to be sharp. They easily cut through R-19 foil or kraft-paper-faced batts even when they're fully lofted. You don't have to compress the insulation to cut it, as you would if you used a utility knife. ***—Ron Smith, Sunol, Calif.***

Another cutting idea

My partner and I prefer a different tool to cut fiberglass insulation—a machete. We bought ours for about $8 at the local hardware store. The cutting is done with the rounded part of the blade near the tip, with long, light pull strokes. Because it cuts cleanly, drifting bits of fiberglass dust are kept to a minimum. The tool has to be sharp, but it's easy to put a keen edge on the blade with a flat mill-bastard file.

—George Hines and Brent Spohn, Ganges, B. C.

Insulating a floor

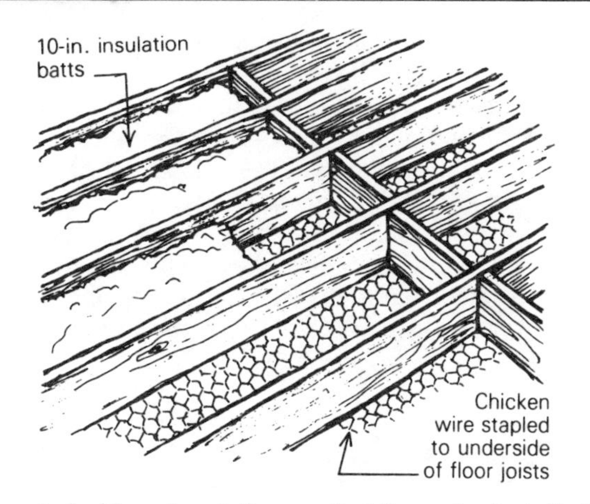

One way to hold up insulation so that it can be installed easily from above is to use twine. But here's an alternative method we used to support 10-in. friction-fit batts in the floor of an addition with a 2-ft. crawl space. First we stapled light-gauge chicken wire to the underside of the floor joists, using 9/16-in. staples. We then dropped in the batts, installed a plywood subfloor and continued framing. The insulation process is fast and easy, with little discomfort to the worker. And the insulation "breathes" well with standard venting.

—***Doug Jackson, Mission, B. C.***

Removing fiberglass slivers

In the course of various home maintenance and construction tasks, I sometimes have a brush with fiberglass when I'm not wearing gloves. Such encounters always leave microscopic slivers in my hands. These slivers are too small to see and to grab with tweezers, but definitely not too small to cause pain and irritation at the lightest touch.

For many years I have used a remedy that works nearly every time. I spread some white wood glue on the area where the slivers are hidden, allow it to dry, then peel it off in a smooth slow motion. The type of glue is important; it must form a tough yet flexible skin that holds together when it is peeled off. Most of the time the slivers are lifted off with it. On those occasions when they aren't all removed, a second peeling in the opposite direction will usually do the trick.

—***Joseph Dawes, Big Springs, Tex.***

Form-fit vapor barrier

When I install a plastic vapor barrier over a wall or ceiling, I ignore the electric outlet boxes until I'm finished hanging the sheet. Then, rather than cut the plastic on the outside of the box, I make the cut ¼ in. to ½ in. inside the perimeter of the box. I stretch the plastic around the outside, which gives me a nice tight fit that cuts down on cold-air infiltration.

This trick works well for other utilities that poke through the vapor barrier, such as ductwork or plumbing. A bead of caulk or duct-tape collar can make the seal even tighter.

—Susan Caust Farrell, Searsport, Maine

Termite protection

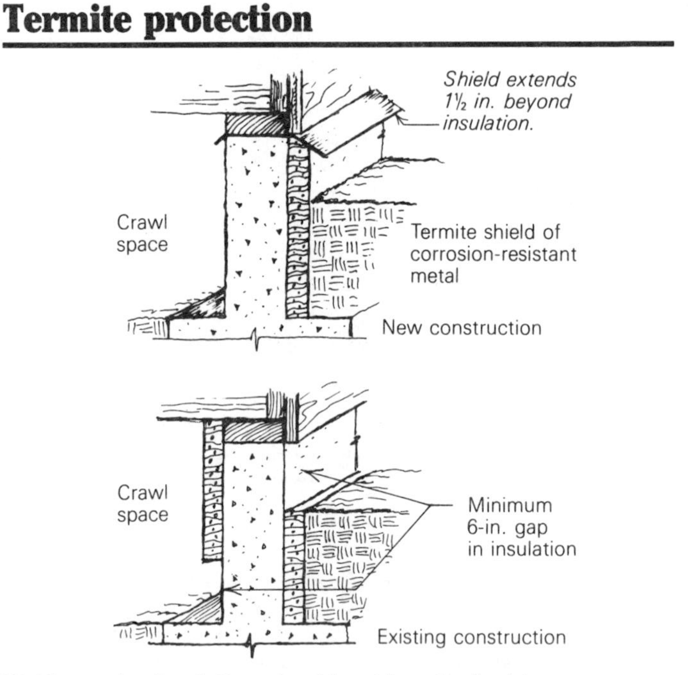

Rigid exterior insulation should not be attached to foundation walls (as commonly recommended in passive solar manuals) in any area where subterranean termites are found. It gives termites hidden access to the walls of a building. In new construction, an unbroken through-the-wall termite shield should be installed as shown in the drawing. For existing construction, either insulate inside the foundation wall (but not to the ground) or stop exterior insulation at least 6 in. below the top of the foundation. The

6-in. space will allow you to see and destroy any termite tubes constructed between the ground and a wood structure. It's a lot better to lose a few BTUs than the framework of your house.

—Tom Bender, Nehalem, Ore.

Smoking out air currents

Air currents are thought to be too unpredictable to be individually described or engineered. For the past five years, however, I have been exploring the convection flow within houses. Though I've used chart recorders and thermocouples to study currents, one simple but effective indicator of air movement is a lighted cigarette.

Drafts coming into a house are apparent because we can feel them, but air leaving the house is harder to detect and remedy. By lighting a cigarette and walking slowly around a room, you'll know immediately where you're losing heat by the direction of the smoke. You can also gauge the relative effectiveness of your window glazing this way. If the smoke drops sharply in front of the glass, you are witnessing sheeting, a downward convection or flow due to the cold your glazing is conducting. The faster the downward flow, the lower the R-value of the window. The smoke is also useful for determining the general circulation of heat throughout your house; by opening or shutting doors, you can get a good idea how most effectively to control the heat within it. Air is really very orderly, and we should consider its movements within buildings when designing them.

—Philip F. Henshaw, Denver, Colo.

Low-budget insulation

Insulation saves money in the long run, but costs plenty when it's first installed. Since I couldn't afford to insulate my entire house when I built it, I did the walls first. Now I'm insulating the ceiling—for free.

I'm filling the spaces between my ceiling joists with the foam "peanuts" used as packing material around fragile goods. I asked several local businesses that receive such shipments to save the stuff for me, and they've been more than willing. I'm amazed at how fast the piles add up.

—Chris Thyrring, Halcyon, Calif.

Airflow gaskets

Last summer I put gaskets around our electrical outlets to cut down on air infiltration, and I recycled our egg cartons at the same time. I bought the first gasket, then I used it as a pattern to mark the outline on the lids of our plastic-foam egg cartons for all the rest. The recycled material makes a good seal. ***—Jackson Clark, Lawrence, Kansas***

Perimeter insulation detail

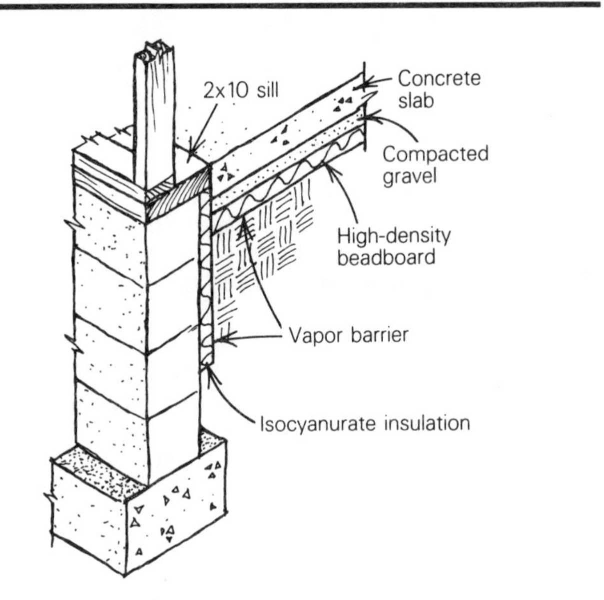

The drawing above shows a detail that we (at the Colorado Mountain College Solar Program) have developed to insulate foundations in cold climates. Although it uses a standard 8-in. wide concrete-block wall, our method differs from typical foundation treatments in several ways. First, we top the block wall with a redwood or pressure-treated 2x10 sill. Its anchor bolts are recessed so they don't protrude above the sill, and the sill overlaps the blocks on the interior side, where it protects the top edge of an isocyanurate rigid insulation panel. This insulation has a higher R-value than extruded polystyrene, but it is more fragile and it can lose its effectiveness if it gets wet. By placing it on the inside of the foundation and protecting it with a vapor barrier, we avoid these pitfalls and keep the slab thermally isolated from the foundation. This detail also eliminates the extra hassle of flashing and finishing exterior insulation, sealing the gap between the foundation and the typical sill and the kind of insulation damage that can occur during backfill. When the framed wall is lifted into place, the recessed anchor bolts don't interfere with the bottom plate.

If the slab is going to be a direct-gain heat sink or contain a radiant heating system, we place high-density beadboard under the slab for better performance. When it comes time to pour the slab, the wide sill plate serves as a screed, and if the floor is to be carpeted, it's a convenient anchorage for carpet strips.

To build with this foundation detail you have to alter the typical construction sequence because the sill has to be installed and leveled before the slab concrete arrives. It also costs extra for the 2x10 sills, but we think that the advantages outweigh the drawbacks.

—Johnny Weiss, Glenwood Springs, Colo.

Chapter 12: Drywall

Bending drywall

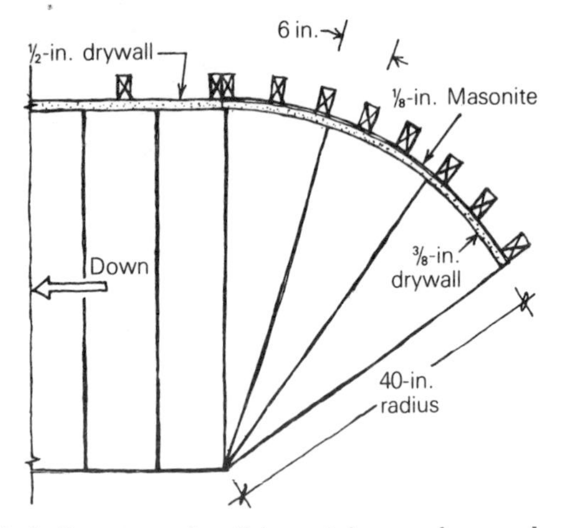

I recently built a curved wall in a stairway of a new home. The wall and top three treads needed to turn to avoid a chimney and to make a smooth approach to the top of the stairs from any of the upstairs rooms.

The curve was a 4-ft. section of a 40-in. radius, framed with 2x4s, 6 in. on center—too tight a curve to bend ½-in.

gypboard without first making an elaborate bending jig. Quarter-inch drywall seemed too fragile for such a heavy traffic area, so I chose ⅜-in. board with an added advantage. I first applied ⅛-in. untempered Masonite to the framing, with widely spaced drywall screws. This took up any irregularities in the framing and made a very smooth curve for applying the ⅜-in. drywall. To facilitate bending the gypsum, I laid it face up on a dew-drenched lawn for about 20 minutes. I then screwed it to the studs and Masonite, starting along one edge, squaring it to the wall and ceiling. Putting pressure against the sheet well ahead of where you are screwing, and working from the center toward each end of successive studs, reduces the chances of buckling. The result was a smooth, very firm, puncture-proof wall. *—Craig Schoppe, Milbridge, Maine*

Corner patch

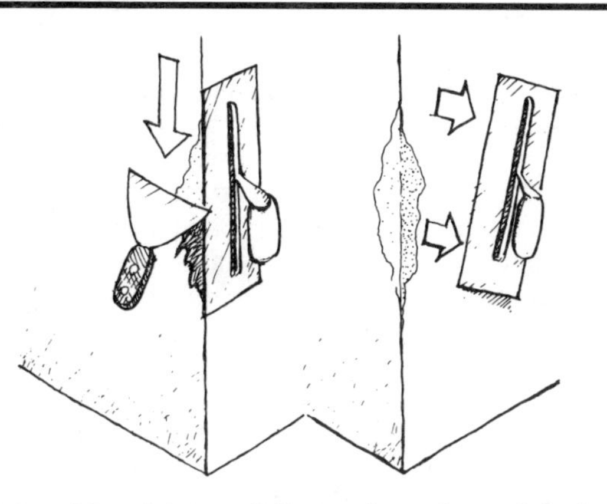

A frequent problem in remodeling and repair work is the damaged outside corner where some plaster has chipped off, but not enough to warrant installing a corner bead. To fix these dings, I take a flat plasterer's trowel (ideally longer than the chipped area) and lay it on one side of the corner, flush to the edge where it acts as a form. Then I fill the exposed side of the hole with patching compound, and slide the trowel away from the corner without lifting it off the wall, as shown above. For larger fill-ins, plaster works better because it doesn't sag as much as patching compound. When using plaster, be sure to apply water or a bonding agent to the old surface.

—Sam Yoder, Cambridge, Mass.

A bending form for drywall

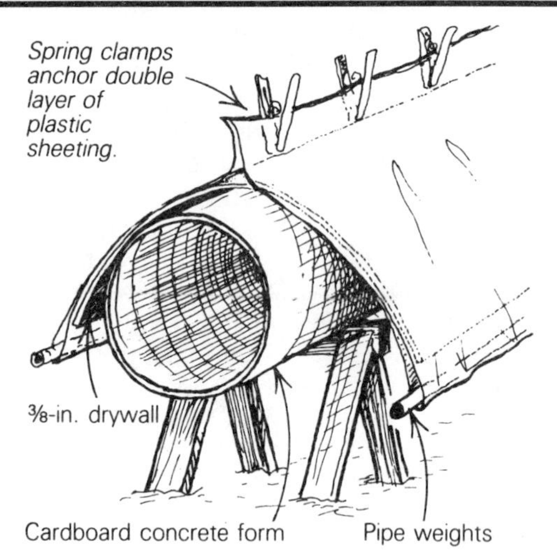

When I had to bend some ⅜-in. drywall into sections for a coved ceiling, I hunted around for some readily available materials that could serve as a bending form and a press to apply even pressure to the panels. I found a 10-ft. long, 20-in. dia. concrete column form, and propped it up outdoors on sawhorses. This became my bending form. I centered a piece of drywall on it, finished side down, and sprayed water on its back. Then I arranged a heavy plastic tarp over the panel, as shown in the drawing above. I doubled the plastic so that it could carry ¾-in. steel water pipes, which hung down at the sides of the form and applied the necessary pressure to persuade the drywall into its arc. Where the edges of the plastic came together, I held them fast with spring clamps.

I had to wet the panels every few hours to get them to bend, but bend they did. When they took on the correct curve, I pulled away the plastic and let them dry until stiff.

I installed the curved pieces of drywall in a double layer, using panel adhesive to laminate them. The resulting cove didn't need curved wooden backing, and it took very little mud to feather the separate pieces into a continuous curved surface. *—Bruce Misfeldt, St. Louis, Mo.*

Folding drywall

Many of today's space-conscious designers have all but eliminated single-flight stairs. Instead they favor stairs that are

U-shaped in plan, with two flights and a landing, because they take up less area. This does indeed save space, but it makes it impossible to carry an 8-ft. sheet of drywall upstairs.

An old builder I know suggested that I score the backside of the sheet at the 4-ft. mark, fold the two front faces together gently and then take the doubled sheet up the stairs, unfold it and screw it in place. This works like a charm, and the extra taping involved is negligible. Most 12-ft. sheets will negotiate stairs this way as well. ***—Art McAfee, Edmonton, Alta.***

Screw-bit encore

To get more life out of drywall screw bits, file the tip when the sides become rounded over. This allows the bit to enter the screw head a little deeper, thus engaging a part of the bit that still has crisp edges.

—Mick Cappelletti, Newcastle, Maine

Patching drywall

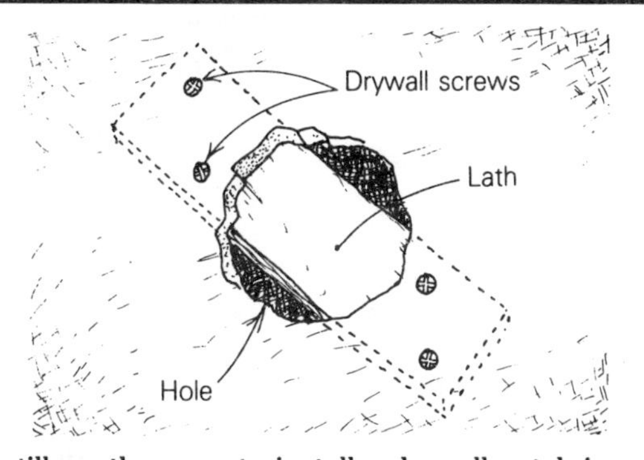

Here is still another way to install a drywall patch in a spot that doesn't have any backing. First, cut an oversize patch, and nibble away at it until it fits the hole. Next, find a piece of plywood or lath that is longer but narrower than the hole. Slip the wood behind the drywall, and position it so that it spans the hole, as shown in the drawing below. Drive drywall screws through the wallboard and into the lath to draw it up tight to the back of the gypboard. Screw the patch to the lath and you're ready to fill the cracks with joint compound.

—William Barstow, Arcata, Calif.

Drywalling at an angle

When we remodeled our house, we used a drywall trick that eliminated taping the obtuse angles formed by walls and sloped ceilings. Instead of creating a joint at this line, we scored the paper on the back and broke the gypsum so that only the face paper remained intact. We kept this hinged piece narrow, about 2 in. on either side of the break. Nailing up this sheet was a bit awkward, but it gave us flat-taped joints on either side that we were able to deal with easily.

—Steve and Becky Benton, Des Moines, Iowa

Drywall battens

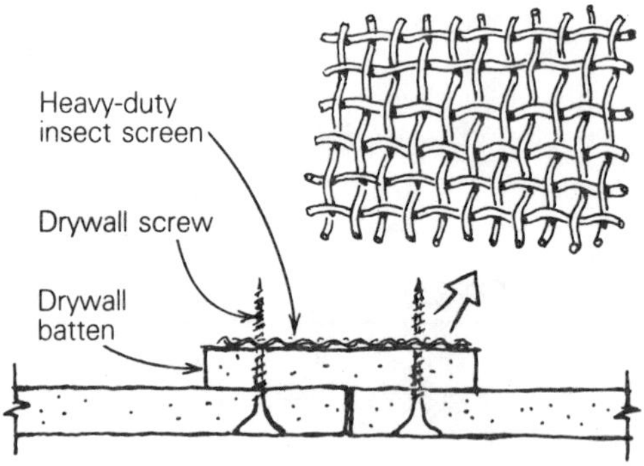

When it's impossible to back the end joints of drywall on a ceiling, I use a batten of drywall about 6 in. wide, as shown above. I screw through both layers with drywall screws, and secure the screws with heavy-duty insect screen. When the sharp screw tip hits the insect screen it finds the nearest void in the weave, spreading the strands apart until the shank can pass through the screen. The threads bear on the wire strands, giving me a solid purchase for the screws.

—Robert L. Kennedy, Grand Forks, N. Dak.

Trimming outlet boxes

Despite everyone's best intentions, switch boxes and electric-outlet boxes sometimes project a bit beyond the plane of the drywall. Left this way, the cover plate won't seat properly, leaving an unsightly gap around the plate. To eliminate this problem with plastic and fiberglass boxes, I grind down any protrusions with a wood rasp. This operation is best done right after the drywall is installed—before the electrician arrives.

—***Sebastian Eggert, Port Townsend, Wash.***

Drywall lift

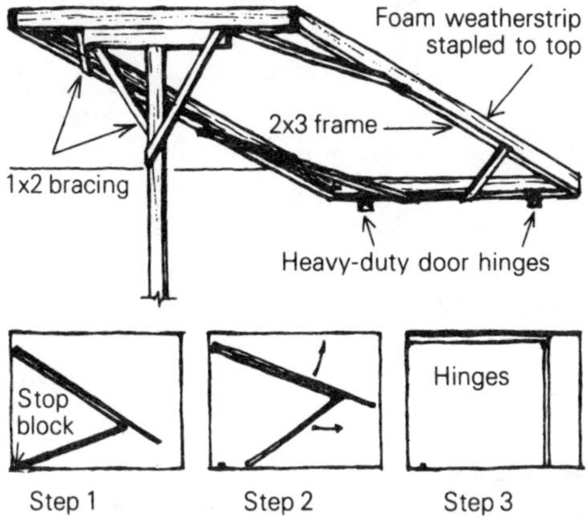

While building a recent addition I used the lift shown in the drawing to help position 4x12 drywall panels on the ceiling. The lift consists of a rectangular frame with a T-shaped leg hinged to one end. I built the frame and leg from 2x3 fir and braced it diagonally with 1x2 strips. My version is 3¾ ft. wide and 10 ft. long. The length of the leg depends on the ceiling height.

I hinged the frame to the wall, allowing about ½-in. clearance between the ceiling joists and the panel when the frame is lifted into position. My hinge wall was a former exterior wall sheathed with wood siding, on which I could attach the hinges at any position. An all-purpose version of the lift should include an additional horizontal member, permanently hinged to the frame, which could be fastened to the wall studs.

For loading, the frame is positioned to allow the panels to clear the floor and secured with a stop block temporarily nailed to the floor behind the leg (step 1 in the drawing). I stapled foam weatherstripping to the top of the frame to cushion the drywall and prevent slippage when the assembly is pivoted toward the ceiling (step 2). Once horizontal, the frame is held in place with the now vertical leg (step 3).

—***Paul Turnrose, Forestville, Conn.***

Drywall splits

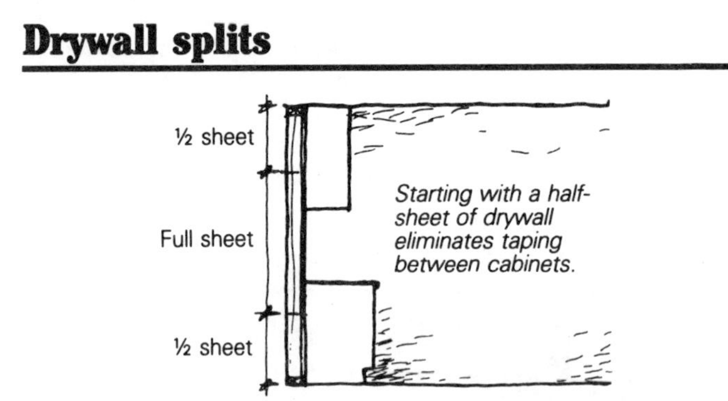

In kitchen areas, where people like smooth painted walls or wallpaper, I cut the first sheet of drywall in half lengthwise, as shown in the drawing, to keep a seam from showing up between the upper and lower cabinets. This does result in an extra drywall seam, but since the cabinets will cover the joints, they only have to be fire-taped.

—Jim Blodgett, Roy, Wash.

Arm extension

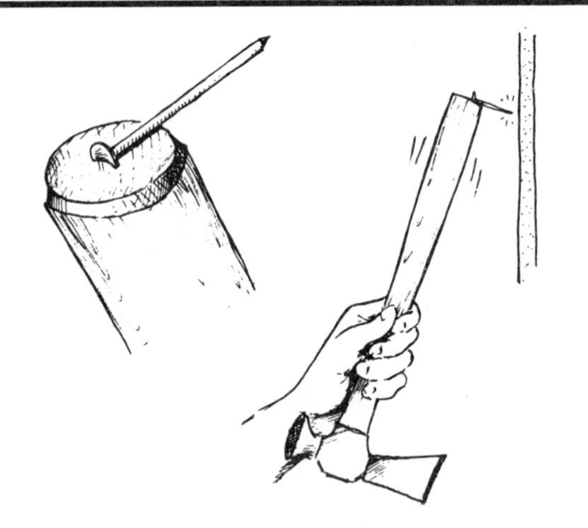

Sometimes when I hang drywall, I have to tack down a spot that's about 10 in. beyond my reach. Rather than move the scaffold to drive one nail, I use my wooden hammer handle to extend my range. I rest a nail on a hard surface, and drive the handle butt down on the head of the nail until it's embedded in the handle, as shown above. Then I hold the hammer upside-down, set the nail where I want it, bring the handle back into my hand and drive it home. This trick also works with some hammers that have handles covered with rubber sleeves. Just tuck the nail head into the small hole at the center of the handle butt.

—Dell Wade, Seattle, Wash.

Errant drywall nails

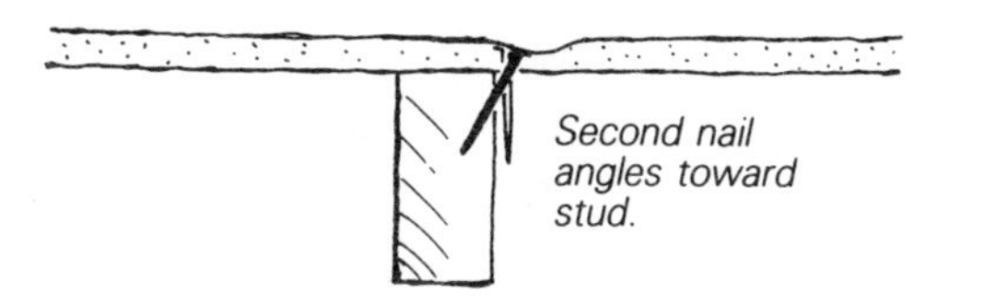

Because of a crooked chalkline or a bad guess, I sometimes miss the stud or joist as I nail off the drywall. I used to remove these nails by plucking them out of the wall with my fingers, which is an annoying task, or I'd drive them through (the sloppy solution). Then a friend showed me this simple trick: drive another nail right next to the misplaced one, but angle it toward the framing, as shown in the drawing above. When this nail is set it will pull the dud down tight, under its head. This takes little time, and the resulting dimple can be finished normally with mud.

—Kurt Lavenson, Berkeley, Calif.

Removing drywall

A short piece of copper tubing is an excellent tool for removing drywall. First locate the nails, either by prying loose one edge of the drywall panel or by tapping your way down the studs. Then place the end of the tubing over the nailhead and give it a whack with a hammer. One hit is usually sufficient to isolate the nail from the drywall. Repeat this procedure for the other nails in the sheet, and it should pull off the wall with no struggle.

—Peter Lewis, San Francisco, Calif.

Coving ceiling drywall

Coved ceilings used to be common in houses with lath and plaster, but now in the drywall era they're pretty rare. I think the aesthetics of some rooms can be greatly improved by graceful curves at the junction of wall and ceiling, and this is how I make them with gypsum wallboard. First, I cut backing blocks with the desired inside curve from scrap 2x material. I nail one to each rafter or joist at the top plate, as shown in the drawing below. The wall studs need not line up with the blocks.

To prepare the drywall for hanging, I score and break the backside of each sheet lengthwise along a series of evenly spaced lines in the area where the curve will occur. I do this carefully to avoid breaking the paper on the finish side. The width of the segments depends on the radius of curvature; for a 6-in. radius curve, I break a segment at about 1-in. intervals. Scoring and breaking must be done one line at a time, but it's easy with one person at each end of a sheet, sandwiching the drywall between two lengths of 2x4 to make each break.

Start with one side of the ceiling and work your way across to the other cove before doing the walls. Nail up the ceiling portion of each coved sheet first, then push the wall portion upward to force the segments against the backing blocks. The result should be a smooth, rolltop-desk effect. The segmented curve can be easily faired in with joint compound. I do this by using light downward strokes with a wide flat trowel and finish with a horizontal pass using a piece of sheet metal bent to the curve.

—Steven M. White, Berkeley, Calif.

Cork patch

Holes in drywall can be patched using corks. Excavate the hole until it's slightly smaller than the cork, and tap in the cork until it's tightly compressed against the drywall. Tapered corks work best. Now trim it back with a knife so that the cork is slightly below the drywall surface, and fill the dimple with drywall compound or Spackle.

—Frank Bonadio, Rochester, N. Y.

Another patch

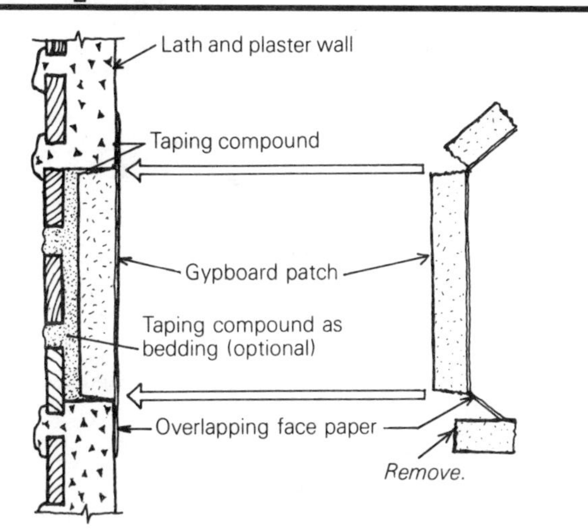

A simple and effective way of patching small holes in drywall or plaster is done without tape. First, cut out the hole until you've got a rectangle. Then cut a piece of drywall of any thickness to the size of the hole plus 2 in. in both length and width. On the back of this patch, score and break the board to the exact dimensions of the hole, and peel away the paper and gypsum from the front layer of paper. You will be left with a solid piece of drywall the size of your hole, with 1-in. strips of face paper on all four sides.

Now apply taping compound to the edges of the hole, and press your patch into the hole until it's flush with the rest of the wall. The face paper acts like tape, and you finish as you would any other drywall joint. If you cut your patch for a tight fit, you can use this technique on the ceiling.

—Geoff Parkinson, Vancouver, B. C.

Utility-knife tune-up

When I work with drywall, I find that the blade of my retractable-blade utility knife cakes up with gypsum dust, rendering the tool useless. To give the blade a longer working life, take the handle apart and run the point of a soft graphite pencil over the blade and its guides. This will remove any built-up gypsum particles, and at the same time deposit a layer of lubricating graphite that won't attract additional gypsum dust.

—Paul J. McCarthy, Hartford, Conn.

Another drywall lift

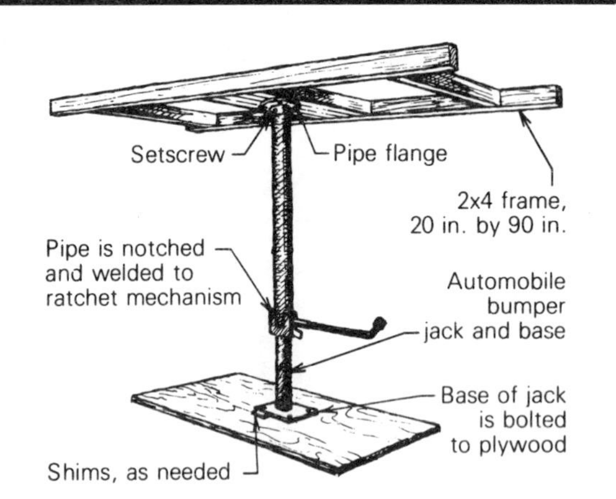

My homemade drywall hanger, basically a frame that can be raised up with an automobile jack, takes a lot of the work out of putting drywall on a ceiling. The size can vary—mine has a 24-in. by 48-in. plywood base with a 20-in. by 90-in. top, which easily accommodates 4-ft. by 12-ft. sheets of drywall. The unit can be jacked up to a height of 8½ ft. The riser pipe should be of a large enough diameter to slide over the jackstand. Notch and weld one end to the ratchet mechanism and screw the other end into a flange at the top. Many automobile jacks aren't made to stand plumb, so some shimming of the base may be necessary.

The completed unit is surprisingly stable and should be able to accept 4-ft. by 16-ft. drywall sheets with no trouble at all. The jig is portable and easily disassembled into three pieces for storage by unscrewing the top and lifting the jack out of the base.

The lowered height of my unit is about 6 ft. so my wife and I can easily left a sheet overhead and slide it on top of the carpet-padded lifter. Then it is simply a matter of jacking it up into place. Our old house has very few square corners, and the sheet can be trimmed to fit while up in the air. There is enough play in the jack to allow a back and forth movement of about 4 in., for fine tuning.

—John Bower, Lafayette, Ind.

Rocking over receptacles

Switch boxes at 4 ft. o.c. can be cut out in place.

T-square

Mark outlet coordinates on subfloor.

Whenever I have to install drywall I mark the location of electric receptacles with a 4-ft. gypboard T-square. I set the square on the floor, T down as shown in the drawing, and I mark the position of the receptacle sides on the subfloor. I also measure the top and bottom edges of the box and write these numbers down on the floor next to the side marks. I do this throughout the room before any wallboard goes up. The only exceptions are receptacles that will fall at joints between sheets of drywall, which can be cut out in place.

Then I put up my drywall right over the boxes and tack the sheets in place. Next, I transfer the coordinates I wrote on the floor earlier to the tacked-up drywall, cut out the receptacle openings with a utility knife, and nail off the panels completely. ***—Ed Kroger, Cincinnati, Ohio***

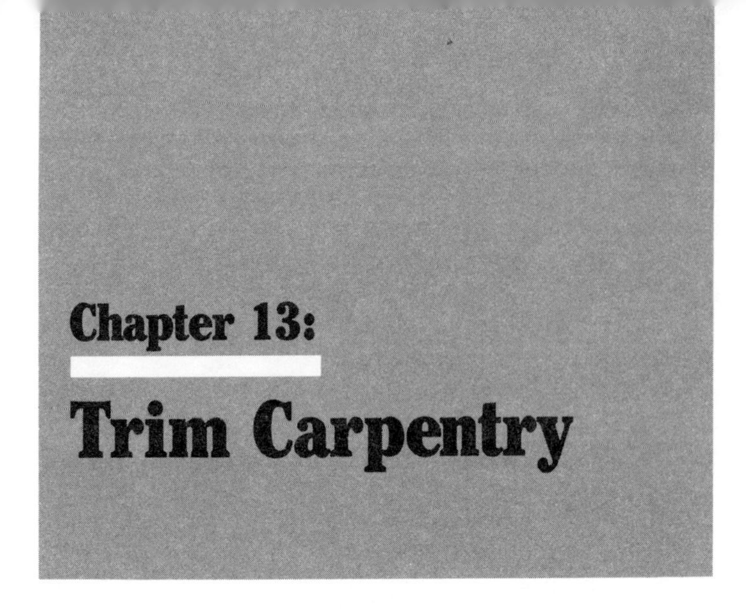

Cutting miters

When casing windows or doors where miters are used, anyone who has done a few knows that they are never square, and often not flush with the wall surface. Rather than using a block plane or rasp to compensate, I've learned it's quicker and more efficient to use the power miter first.

Cut the first piece (top or bottom) to length with true 45° angles (miter boxes I've seen have adjustments to true the cut and should be checked periodically), and nail it in place with an even reveal of the jamb.

For the sides, start by cutting a true 45° (90° to the back surface), then hold the piece in place with a consistent reveal, tight to the wall and jamb. If the joint is open at one end of the cut, take it back to the saw and compensate by changing the angle of the material (not the saw) until you get the right cut. If it is still open—but evenly so—across the front, back-cutting is required. To do this, simply insert a thin wood shim to get a small trim angle. Once you have that fit, hold in place at the opening, mark the length, and cut it at a true 45°. (If you are casing a door, just cut the bottom end off square.) Repeat for the other side.

To fit a final piece where two ends must come together, use the same method. However, you must leave the piece long, so both ends can be trimmed to fit. Once you have

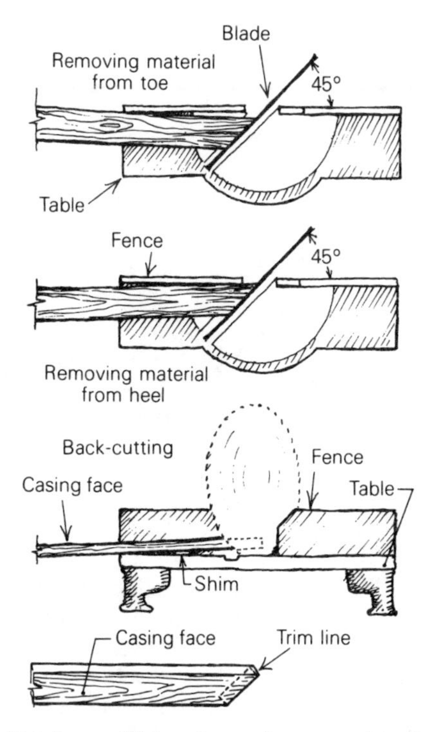

both ends fit (piece still long), mark one end to final length and repeat the fitting cut.

This is a safe and—with a little practice—easy way to get clean fits every time, quickly. A carbide blade helps, but be sure it can withstand the high rpm of power miter boxes.

—Chuck Sjodin, Coon Rapids, Minn.

Trim advice

Whenever I trim out a door or a window, I first attach the head casing, cut to the finished length. Then I make the miter cut on each leg (side casing) and check the miters for fit. At this point each leg is still about ½ in. too long. Instead of trying to measure the exact length, I turn the trim upside-down, with the point of the miter resting on the floor (or window stool). Making sure the trim is lined up with my reveal marks, I strike a line on the trim even with the top of the head casing. Then I cut the leg square, leaving the line, andand the trim fits tightly every time. This method is helpful to apprentices because they can get clean and accurate results fast, while learning how to use their tools.

—Patrick D. Rabbitt, Greenwich, Conn.

Offset jig

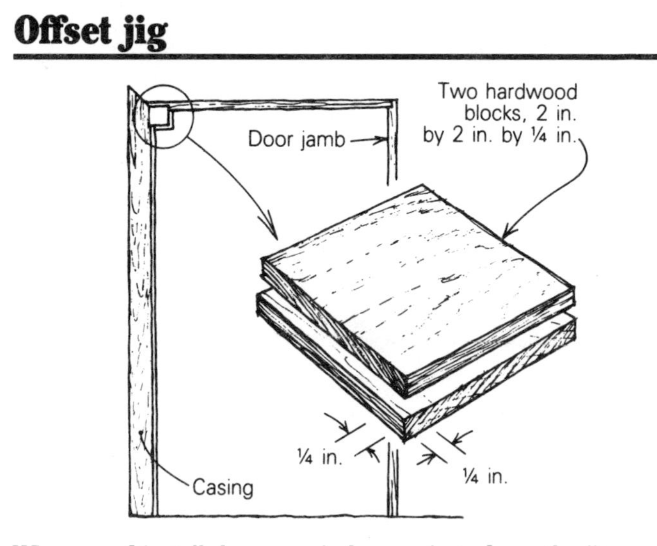

Whenever I install door or window casings, I use the jig shown above to make sure that I get an accurate ¼-in. reveal. It's made of two square pieces of hardwood, ¼ in. thick, that are glued together with a ¼-in. offset. This jig has so many corners that half the time I grab it out of my nail bag it's in the right position for use.

—John Sandstrom, Fort Dodge, Iowa

Barnboard basics

I recently installed some weathered boards taken from an old barn as wainscoting in our dining room. In doing so, I used several tricks that may be helpful to others who want to use this rustic but fragile material.

First, nail it to a solid underlayment. Although furring strips will work, a layer of ½-in. plywood is better. It lets you tighten up a board with a nail wherever it's needed. I painted the plywood a medium grey, which minimized the impact of empty knotholes and the cracks between the boards. I also used a spray can of grey paint to touch up the ends of freshly cut boards. Left unpainted, these ends have a nasty way of showing.

Cupped boards can be trouble when you're using weathered material for paneling. I got around this problem by making two or three saw kerfs down the length of each board. The cuts come to within about ⅛ in. of the good side, and they allow you to bend the cup out of the boards as they are nailed to the wall. You can expect to waste a few boards with this method, but the ones that survive will give the desired rustic effect without looking shabby.

—Dale Nichols, Toledo, Ore.

Trimming round corners

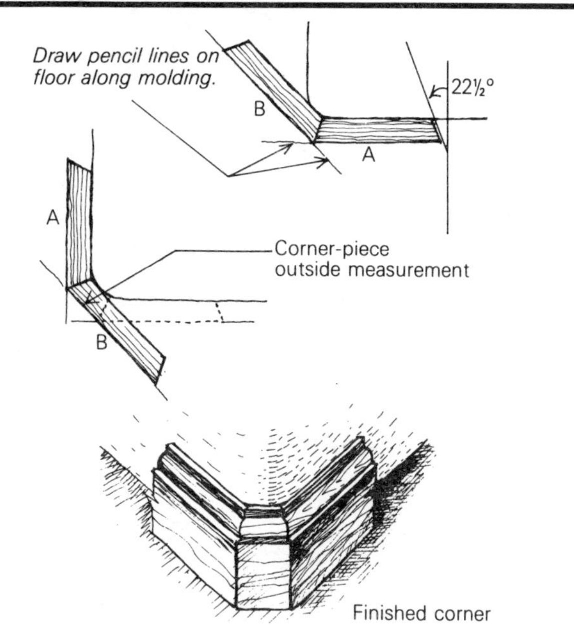

The following procedure allows you to install baseboard or crown molding where the plaster or stucco corners are radiused, without leaving an ugly gap behind the heel of the miter. Begin by cutting two pieces of base molding about 4 in. long with an angle of 22½° (one-half of 45°) on each end, as shown in the drawing above. Then, holding piece A flat against the wall and piece B against the round corner, move both pieces until a tight corner fit is formed. Mark this junction with a pencil line on the floor. Next put piece A flat against the other wall and move piece B along the pencil line until again a tight corner is formed. Now scribe another pencil line along the front of piece A. Where the lines cross is the given measure of the small corner molding piece, which is also angled at 22½°. The long pieces can also be accurately measured from the intersecting lines.

—Hans Matzinger, Laguna Beach, Calif.

Forming curved trim

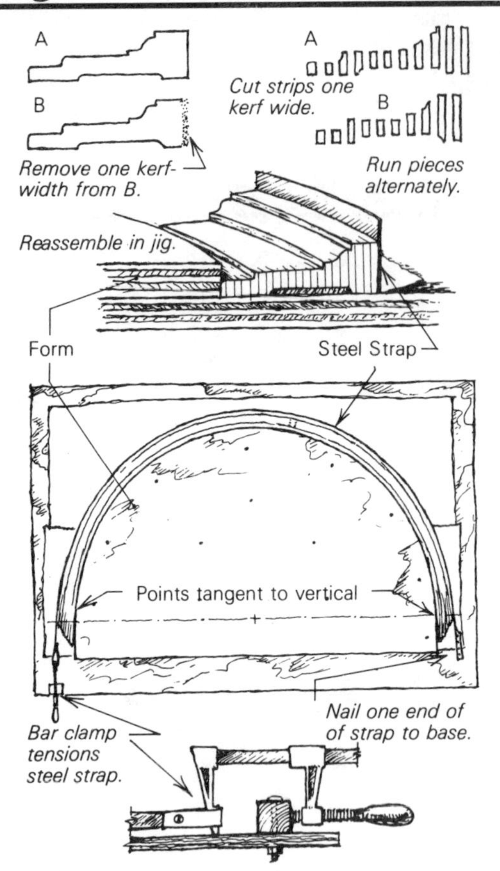

Trimming arched windows can be a problem, especially with complex molding patterns. This method requires two pieces of molding to form one finished piece of curved, paint-grade trim.

First, determine the distance between the inside edges of the vertical moldings (let's say 4 ft.). Cut the appropriate radius out of a sheet of plywood thicker than the trim and nail it to a plywood base. Our 4-ft. diameter arch will need a 2¾-in. trim piece a little over 7 ft. long, cut from two 8-ft. pieces (A and B in the drawing) to allow a margin of error.

Remove material the width of the sawblade (the kerf) from the edge of piece B and then set the table-saw fence to leave a strip exactly one kerf wide with each piece. You don't have to remove the fence again. Running the two pieces through the saw alternately makes it easier to collate the thin strips. If your molding has a hollow back, tack shims in the jig, one foot on center, to ensure proper alignment during clamping. Baste the strips with your favorite glue, place them in sequence against the curved form, and clamp the steep strap in place. I use a 26-ga. steel strap cut from an old downspout for a clamp; one end is nailed to the base, and the other has a loop large enough to anchor the end of a bar clamp. A bar clamp is better than a C-clamp for this application because the sliding jaw allows tensioning without exhausting the reach of the threaded shaft.

When the glue has set, scrape off the excess and sandpaper the rough edges. The junction with the vertical trim should be a 90° cut in the base of the arch at the point tangent to vertical. ***—Jud Peake, Oakland, Calif.***

Beading tool

Here is an easy way to cut a beaded edge in softwood. For the handle of the tool, use a hardwood scrap about ¾ in. by 1¼ in. by 5 in. Drill a pilot hole in the end of the block, and run in a #10 or #12 flathead wood screw, 2 in. long. Now clamp the handle in a vise, and use a hacksaw to deepen the screw slot, as shown below. This forms a sharp edge on each side of the slot. By running the screw in or out, you can adjust the location of the bead, but make sure to align the slot perpendicular to the surface of the board you're beading.

With the end of the handle acting as a guide, drag the tool down the work several times until you get a nice groove of the proper depth. The screw head cuts a square shoulder on one side, and a bevel on the other. Finish the bead by easing the beveled edge with a block plane, and touch up with sandpaper. *—James Laws, Salisbury, Md.*

Counter trim

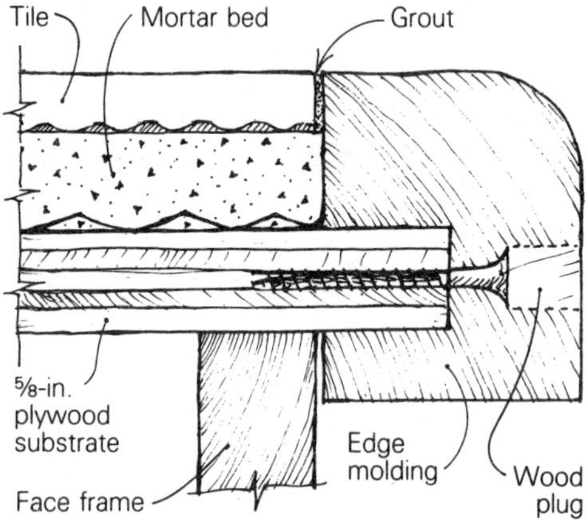

We use the wood counter edge detail shown in the drawing above in our custom home kitchens. We like this edging because it eliminates the need for trim tiles, which can be very expensive and hard to find. In addition, it makes a pleasing accent in a kitchen with painted cabinets. We prefer to use rock maple for this trim because it's stable and hard enough to stand up to daily use.

To make the edging, we start with a 1⅝-in. by 2-in. length of maple, and we rout in a ⅝-in. wide groove to accept the plywood substrate of the counter. The face of the molding can have any profile, but we like to give it a simple roundover—it's nice to lean against and easy to clean. The top of the trim should be at least 1 in. higher than the surface of the rough counter for the mortar base and the tile finish. We bond the molding to the plywood with waterproof epoxy and hold it in place while it sets up with recessed drywall screws.

Before the tile-setter arrives, we stain the trim with a coat of Minwax stain, and rub in a coat of 75% urethane and 25% mineral spirits with steel wool. The trim is masked off during the tile setting and grouting. Then we give the trim a final rubdown of the urethane solution to touch up any blemishes. *—Patrick Miller, Del Mar, Calif.*

Improved V-groove

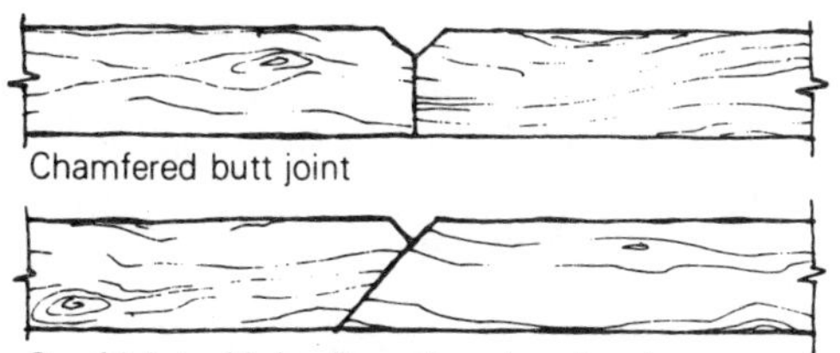

I've come up with a scarf joint for the board ends of paneling that gives a V-groove appearance on the end grain (drawing, above). It has several advantages over a chamfered butt joint. I make an ordinary scarf joint and chamfer the leading edge. This saves the step of the second chamfer, and it conceals the gap between the boards.

—Ezra Auerbach, Lasqueti Island, B. C.

Self-scribing plate rail

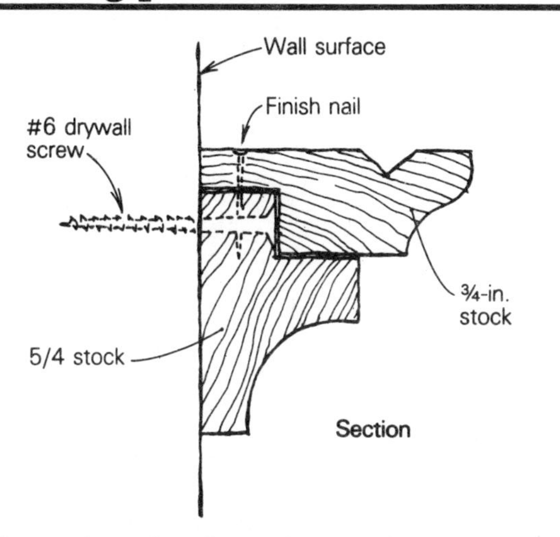

I have clients who collect fancy plates, and they needed more room to display them. Plate rails were the obvious answer, but the only available walls for the rails were already painted, wallpapered and very uneven. I could prefinish each rail to cut down on the risk of making a mess after installation, but the texture of the wallpaper made it difficult to scribe a rail for a good fit.

My solution to this problem was to make up a two-piece rail that didn't require scribing. I attached the coved section to the wall with drywall screws, as shown in the drawing. The screws pulled the molding up to the wall tight enough to eliminate any gaps. Then I nailed the plate-rail molding to the coving from above. The nail holes are hidden from view, and variations in the wall are taken up by the joint between the coving and the plate rail.

—Rod Goettelmann, Vincentown, N. J.

Dressing up a door

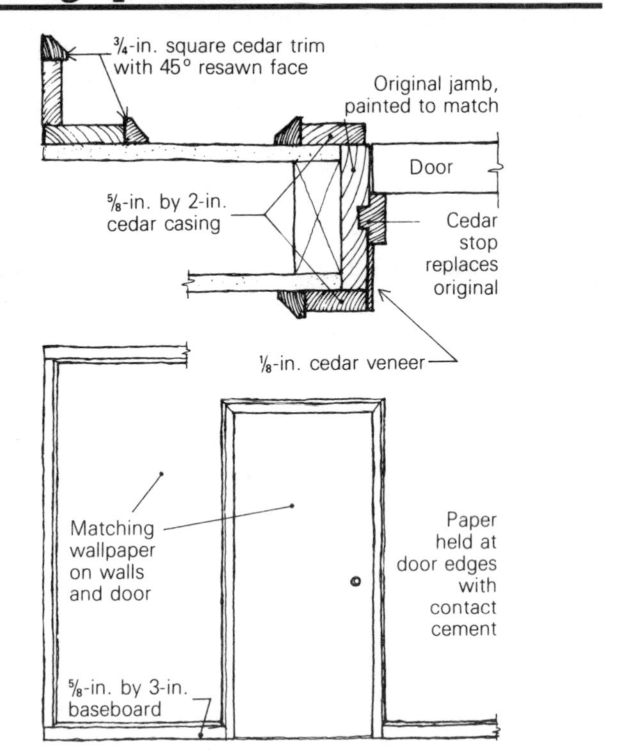

The house I own came with pre-hung doors with rotary-cut skins. The doors were hung in plain mahogany frames. They weren't much to look at, and suitable only for painting. To add some visual interest I cased the door in various thicknesses of cedar trim, as shown, and continued the treatment throughout the room. The ¾-in. bevel-cut trim is slightly darker cedar for added contrast; the original jamb is painted to match. Rather than paint the door, I covered it with the same wallpaper as the adjacent walls, using contact cement to glue the edges down. ***—G. Siguardson, New Westminster, B.C.***

Steam-bending on site

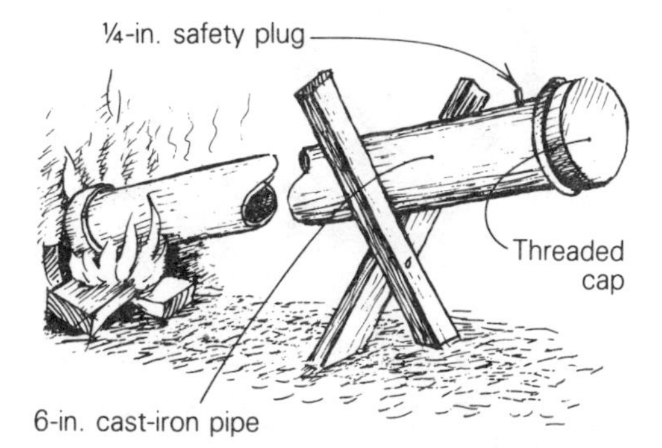

While we were restoring a large Victorian building in San Francisco, we had to bend a lot of redwood trim around 4-ft. and 12-ft. diameter framed partitions and enclosures. Steam-bending seemed like the easiest way to do this, but because the trim pieces were so long, we had to devise an equally long steamer to accommodate them.

I went to a scrap-metal yard and bought a 14-ft. length of 6-in. threaded cast-iron pipe for $25 (8-in. pipe was also available). I also bought two end caps. I put a thin layer of plumber's putty on the threads at one end of the pipe, and tightly screwed on one of the caps. Near the opposite end, I drilled a ¼-in. safety-valve hole, and tapped in a cedar plug. Then we put the capped end into the building's fireplace, built a fire around it with wood offcuts, and poured in about a gallon of water.

About 20 minutes later, the pipe was steaming, and I picked out four pieces of the redwood trim with the most vertical grain, shoved them inside and screwed the second cap on hand-tight. After another 20 minutes we uncapped the steamer, removed the wood (wearing gloves) and cut it quickly. The redwood easily bent to the required curves.

As we grew accustomed to the procedure, we precut the trim pieces because working time is short—about 3½ minutes. Experience also taught us that vertical-grain pieces were much easier to bend than flat-grain ones. With flat grain, the growth rings tend to separate during bending.

The advantage of this steamer is that 20-ft. lengths of trim can be steamed, by joining two pieces of pipe with a union. Also, most job sites have scrap wood and a safe place to build a fire. The disadvantage is that cast iron stains redwood black. This isn't a problem if you're going to paint the trim.

—Michael Spexarth, El Cerrito, Calif.

Cutting double-angle miters

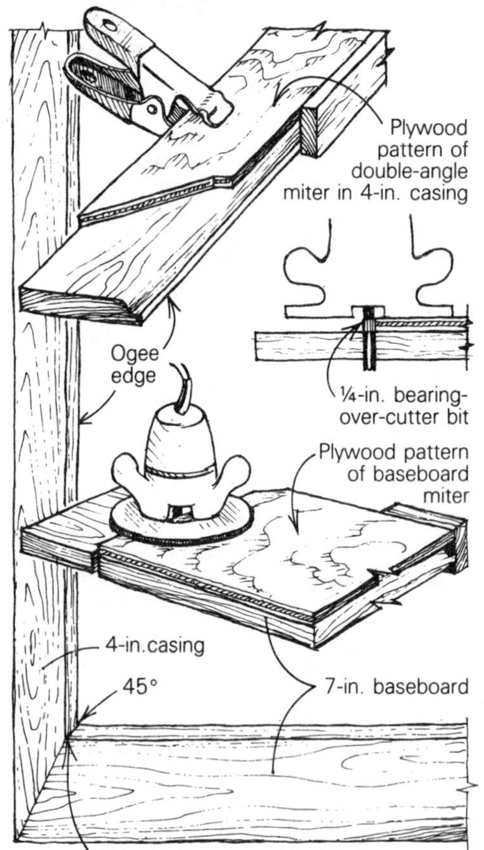

Trim inside miter with chisel.

Recently the general contractor I work for was low bidder on the finish phase of a large Victorian-style office building. The job called for crown moldings, 20 raised-panel doors surrounded by 4-in. casings and hundreds of feet of 7-in. high baseboard. Both the baseboard and the casing had a ⅝-in. Roman ogee milled on one edge. At each doorway, the base and casing were to be mitered. The miter where the ogees met at the top of the joint was 45°, but this angle changed on the flat face of the trim pieces, as shown above. The question was how to cut a lot of these double-angle miters with speed and accuracy.

I solved the problem by drawing the joints full size (both right-hand and left-hand sides) on ⅜-in. plywood, and then cutting each one apart along the lines of intersection. I added stops along the ogee-edge side of each pattern, making it easy to align them with the stock. Clamping the stock and patterns together as shown in the drawing, I cut each joint with a small router using a ¼-in. bearing-over-cutter bit. Armed with these guide jigs, the rest of the crew joined me in production-cutting the miters.

—Bob Grace, San Jose, Calif.

Inside measurements

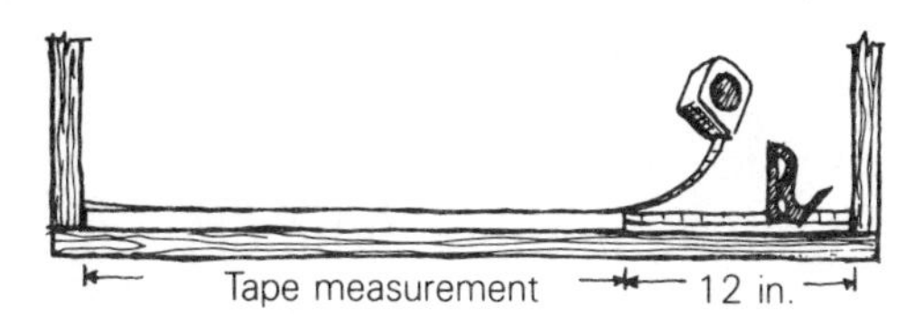

Finish trim work around windows or doors requires accurate inside measurements, but it's hard to get precise readings with a tape measure. The "see through" windows on some tapes don't work very well, and I've never gotten accurate results from measurements that include the length of the tape case.

Instead, I use both my tape measure and my combination square, as shown above. The square's blade butts against the inside surface, while the tape measure records the distance from the opposite edge to the blade. I add 12 in. to the tape reading for the total. This method is fast, precise and makes use of the tools I've already got in my belt.

—Philip Zimmerman, Berkeley, Calif.

Filling nail holes

I've got a system for filling nail holes that I've used for years. It takes a bit of time, but on your best work it's worth it. I start with soft, oil-base jar putty (Color Putty, Monroe, Wis. 53566). This putty is very similar to oil-base glazing compound. I get several shades for each job, including some lighter and darker than what I think I'll be needing. I also get a small can of white, oil-base glazing compound to lighten colors, and some tubes of universal paint tinter (especially raw sienna) to modify the hues. With these materials, I can make all the shades I'll need by mixing and matching.

If the work surface is going to be stained, do it first. For best results, it should have one or more coats of clear finish, sanded lightly. Now mix a generous amount of putty to match the average color of your work. Next, mix some to match the lightest and the darkest colors of the wood. These are the base colors, or mediums. Intermixing these three will provide you with a spectrum of shades, all related to the original bases. For a single project, I may use seven or eight shades—different colors for different parts of the grain—and it sometimes takes several tries to get it right. I fill the holes as flush as possible, and then clean around them with mineral spirits.

On some occasions, when the spring and summer growth rings of the wood grain meet, I've used two different colors back-to-back in the same hole.

—Byron Papa, Shriever, La.

Baseboards for tile floors

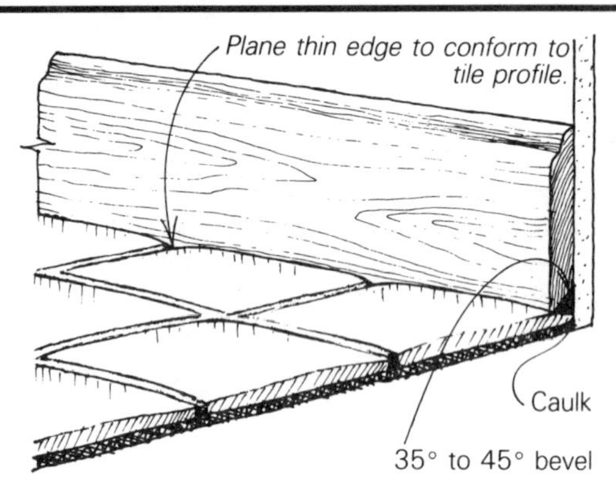

The cost of tile cove base is surprisingly high—about $3 to $4 per running foot. On the other hand, scribing a wooden baseboard to a tile floor is very time-consuming. The method I use to install wooden baseboards gives me a water-resistant joint that looks painstakingly scribed, but isn't.

First, I bevel the back of my baseboard material as shown in the drawing above. The thin edge on the front of the baseboard will usually conform to a well-laid tile floor with a few taps of a hammer—no scribing is necessary. Be sure to use a piece of scrap wood to cushion the hammer blows.

After dry-fitting, I tack the baseboards in place. As I remove them one at a time, I carefully lay a bead of good caulking compound in the void behind the bevel. I like Geocel caulk (Box 398, Elkhart, Ind. 46515). Any caulk that oozes out can be cut away after it sets up. Pre-finishing the baseboards saves time, and usually gives better results than trying to paint or varnish them in place.

—M. Felix Marti, Monroe, Ore.

Baseboard splice

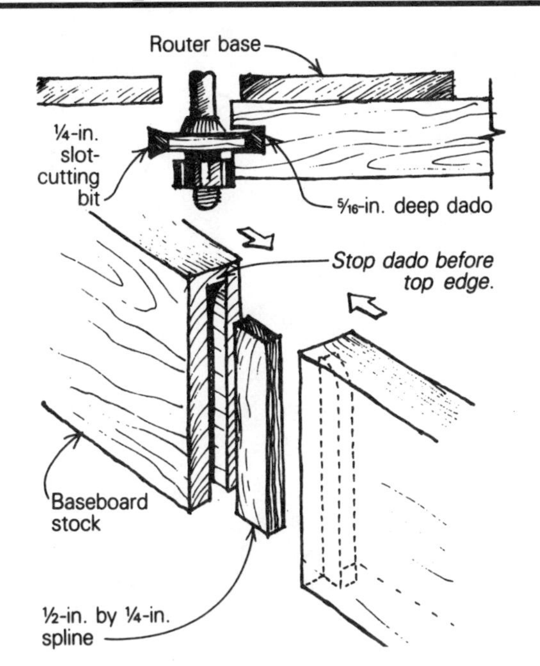

When I need long runs of 1x baseboard, I don't splice my boards with 45° scarf joints. Instead I cut the boards square, letting the breaks fall where they may regardless of stud layout, and I join the pieces with small splines.

I use a ¼-in. slot-cutting bit in my router to cut a 5⁄16-in. deep dado in the end of each board. Care has to be taken here to avoid cutting through the top edge of the baseboard stock. Then I spread glue on a ½-in. by ¼-in. spline that is slightly shorter than the length of the dado, insert it in one of the dadoes, and tap another board onto it. The result is a perfect, tight-fitting no-fuss joint.

With this method, you save time by not having to cut, adjust and recut the mitered joint, and you save material by not having to break the baseboards over a stud. You also avoid the inevitable splitting when you nail mitered joints together without predrilling them. ***—Robert Prasch, Portland, Ore.***

Drilling from two sides

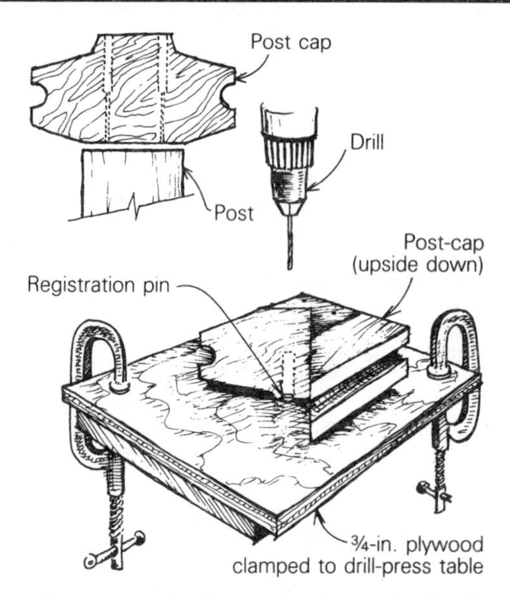

While installing a set of oak posts and guard rails, I ran into a problem: The ornamental post caps were over 4 in. thick and required four screws to attach them to the posts. Each screw had to be countersunk into a ½-in. by 2-in. hole, and because my ¹¹⁄₆₄-in. twist-drill bit was too short to reach through the thick caps, I had to drill them from opposite side. The holes had to be aligned, but the complex shape of the blocks made it hard to transfer the layout from one side to the other.

To solve this problem, I began by drilling the four holes in the top of the post cap. Next, I clamped a piece of ¾-in. plywood to the drill-press table, making sure that the table was perpendicular to the axis of the drill. Then I used my ½-in. bit to drill a hole in the plywood. I plugged this hole with a ½-in. dowel, leaving about ⅛ in. protruding—this became my registration pin.

Without moving the drill or the plywood table, I reinserted the ¹¹⁄₆₄-in. bit in the chuck, turned the post cap over, and one by one, registered the ½-in. holes over the pin. The pin located the exact centerline for each hole, and let me drill the remaining holes quickly and accurately.

*—**Mark Hallock, Arcata, Calif.***

Stain removal

Unfinished wood often gets stained from rusting nails, cement splotches or water leaks. Perhaps you've noticed this kind of staining around the corners of skylight wells that are lined with wood. If you need to remove such stains, try a solution of oxalic acid and water. Mix ¼ lb. of oxalic-acid crystals with 1 gal. of warm water and stir until dissolved. Paint the solution onto the stained areas and let it stand for 15 minutes. Rinse the work with cold water, and repeat if necessary. Oxalic acid is sold at chemical-supply outlets, and it is poisonous. Be sure to wear rubber gloves and eye protection when you use it.

—Ernie Alé, Santa Ana Heights, Calif.

Presanded plugs

When I plug screw holes in places where there isn't room for my sander, I presand the plugs. First I drill some holes in a piece of plywood and I press the plugs into them. Then I sand the plugs smooth, push them back out and tap them flush into their permanent homes. Touch-up is done with a chisel or sandpaper. *—Mark Hallock, Arcata, Calif.*

Hanger-bolt driver

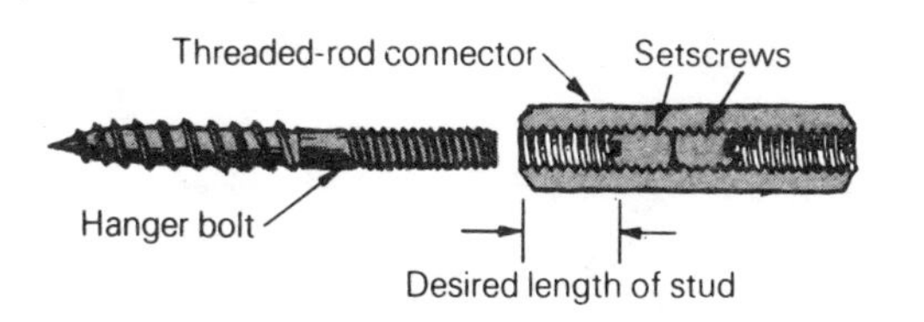

To screw in hanger bolts easily, make up a driver from a threaded rod connector and two setscrews. You can determine the length of the installed stud by how far you drive the setscrews into the rod connector.

Screw the hanger bolt into one end of the connector. Then use a socket and ratchet or electric drill on the other end to drive the bolt into its pilot hole. A touch of parafin on the screw threads will make the job easier.

—Alan Dorr, Chico, Calif.

Splinter-free crosscutting

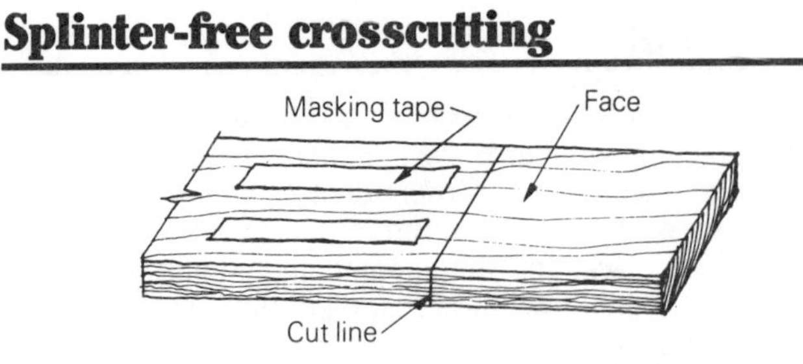

Occasionally in the course of building a home, and sometimes in the shop, it is easiest to use a skill saw to crosscut a piece of trim. With the aid of a saw protractor or some similar device this can be done quickly and accurately.

Usually the face of the work is placed down and the saw base passed over the back of the piece, making a clean cut on the show side. But if it's impossible to cut from the backside, other measures must be taken to avoid damaging the face of the work.

My commercial-duty Rockwell saw has an aluminum foot that leaves black marks on the wood. I've tried polishing the foot with car wax, covering it with duct tape, and using strips of electrical tape to no avail. The best solution so far has been to place two strips of masking tape on the wood, as shown above. Score the cut line with a utility knife to prevent tear-out (this also works with painted surfaces). Cut on the waste side of the line and get as close as possible without touching it. Remember to remove the tape right after cutting to prevent the adhesive from bonding to the wood. ***—Alan Miller, Brush, Colo.***

Bamboo nails

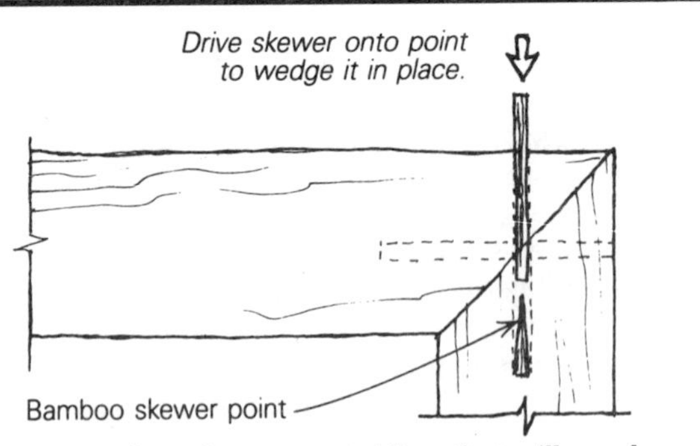

If you have to make a frame or molding that will need planing, carving or routing after it is assembled, consider using bamboo skewers to hold it together as the glue dries. I first started using the skewers to pin picture frames together, as shown in the drawing. First I brush the miters with yellow glue and clamp them together. Then I drill two holes at each corner (one from each side) about 1 in. deep and at slightly different levels so they don't intersect. The holes should be slightly larger in diameter than the skewers. With a pair of wire cutters, I snip off the sharpened point of a skewer. I put a dab of glue on this skewer point, and press it into one of the holes, cut end first. Next I drive another glue-coated skewer into the hole, square end first. The skewer point at the bottom of the hole acts as a wedge to spread the second bamboo pin, locking it in place.

Pins from bamboo skewers are strong and cheap, and you can work the assembled piece without fear of damaging your edge tools or sending dangerous bits of metal flying around the shop as you do with screws or nails. Check your supermarket for these useful fasteners.

—Michael Sweem, Downey, Calif.

Nailset angle

In the course of my work I have to install baseboards from time to time, and I've found that holding the nailset in the manner shown in the drawing above allows me the stability that's necessary for accurate nailing. I wrap two fingers over and two under the set, and I place my weight either on my fingertips and my thumb, or on the heel of my hand.

—Harry Muruaga, Fayetteville, Ark.

Crown-molding fixture

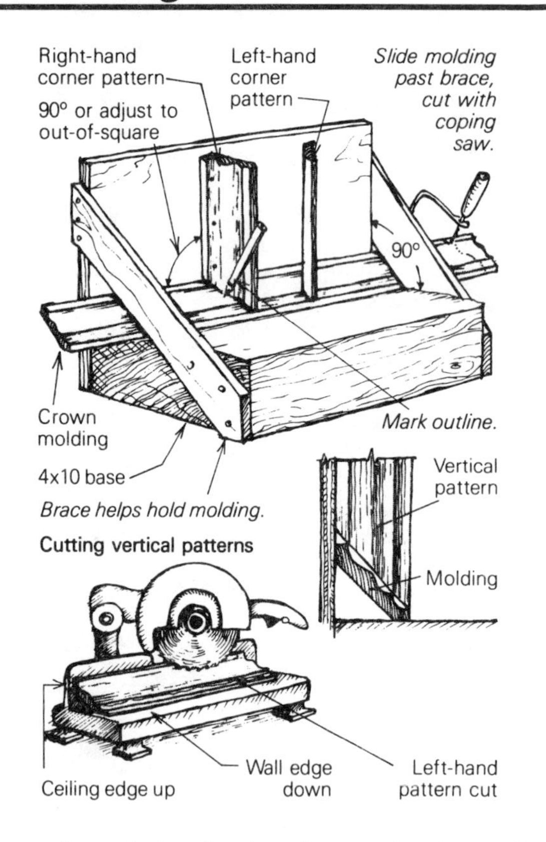

The fixture shown in the drawing above makes it easy for me to mark and cut crown molding for a coped inside corner. I use framing scraps to make the fixture—a 4x10 for a weighty base, ¾-in. plywood for the back, and 1x4s for braces. These braces should be set so that the molding, with its face up, just passes underneath the bottom edge.

For the vertical pattern pieces, cut miters on two short lengths of crown molding, one right-handed, the other, left. These should be cut from stock of the same mill-run that you'll be installing. Nail these two pattern pieces so that they just touch the horizontal molding that is in the fixture for marking. The vertical pattern pieces should be nailed at 90° to the table of the fixture, although they can be adjusted out-of-square if necessary.

To use the fixture, just mark the molding using the right- or left-hand patterns, and slide the scribed end out beyond the fixture for cutting with a coping saw. For marking and cutting, the molding is held securely by the pattern pieces and end braces. A generous undercut and a half-round file for touch-ups will yield a joint good enough to write home about.

—Harry W. Stangel, Palo Alto, Calif.

Crown molding, another way

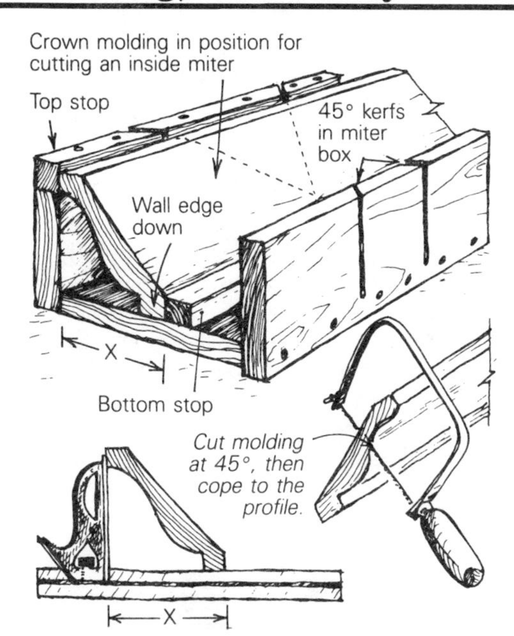

For cutting crown molding I use a fixture made from an ordinary wooden miter box and two 1x stops. First, find the horizontal dimension (X in the drawing above) of the crown molding, then screw the bottom stop in place so that its rear edge is that distance from the back of the miter box, as shown. Cut the other 1x wide enough to support the upper edge of the molding. Screw it in place, making sure that the screws are out of the saw's path.

Cut outside or inside mitered corners by setting the molding against the stops, wall edge down. I think it's more accurate to cope inside corners to the molding's actual profile than to a penciled line, so after I cut the inside 45°, I cope that profile as shown in the drawing.

—Peter Sacks, Somerville, Mass.

Laminate spacer

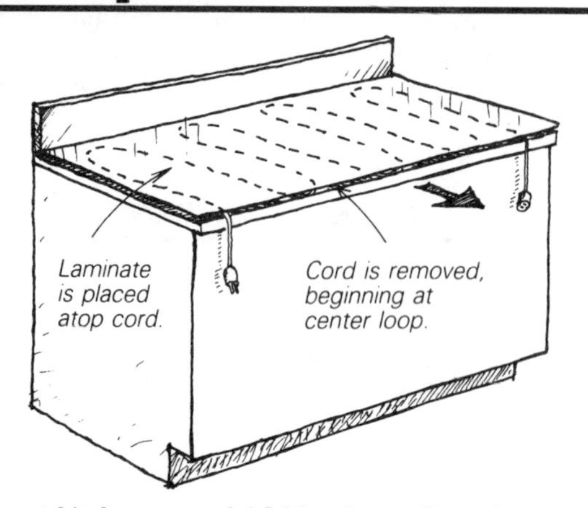

On a recent kitchen remodel I found myself ready to attach the plastic laminate to the countertop with contact cement, but I didn't have my usual stack of spacers that I use to separate the counter and the laminate during alignment. In need of a handy alternative, I turned to a thin extension cord in my kit. I found that by looping the cord across the counter, as shown, I could position the laminate, then remove the cord starting with the loop nearest the center of the counter. ***—Hayes Rutherford, Coolin, Idaho***

Dowel-anchored butt joint

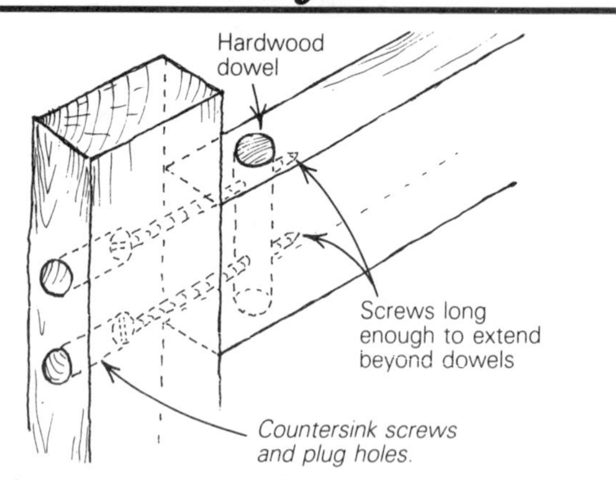

Anyone who spends much time putting together pieces of wood eventually comes across a situation where a simple screw-held butt joint would be both efficient and attractive.

Unfortunately, screwing into the end grain of any wood has little holding power, and glue increases the bond only marginally. Doweling the joint is effective, but tedious. Also, dowels have no ability to pull the joint together, which screws do quite well. The problem boils down to finding a way to keep the screws from stripping out of end grain.

A simple way of overcoming this difficulty is to insert a dowel in the piece of wood receiving the screws. The dowel should be perpendicular to the butt end of the piece of wood, as shown in the drawing, where it will provide a good cross grain anchor for the screw's threads. As a rule of thumb, I use a diameter of dowel that is about half the thickness of the stock. Pilot holes should be drilled through the dowel after it's inserted in the hole to avoid splitting it. Screws should be long enough to extend a bit past the dowel to ensure maximum grip.

—John Grunwald, Hornby Island, B. C.

Searching for studs

While rubbing down some window trim with fine steel wool, I noticed some little black dots forming on the wall nearby. To my amazement, I discovered that some of the steel-wool particles were being attracted to the drywall screws that I had used to install the gypboard three years earlier. Evidently, the magnetic tip of my drywall screwdriver had imparted some of its magnetism to the screws. Now whenever I need to find a stud to hang a shelf or a picture, I just gently tap a piece of fine steel wool against the wall until the black dots magically appear. ***—Jim Murray, Atlantic Beach, N. Y.***

Wood filler

I've found that Bondo and other auto-body fillers like it make excellent wood fillers for exterior and interior painted surfaces. They adhere to wood better than most wood fillers. They don't shrink. They saw, drill and plane better than typical wood fillers. And best of all, auto-body fillers set up in just 30 minutes, no matter how large the cavity you are filling. ***—Jim Stuart, Covina, Calif.***

Sandpaper grip

When sanding a rounded form, such as a hand-rail, finger sanding gives you the best feel for the shape of the work. For a better grip on your sandpaper, use double-sided adhesive tape on the back of the paper. This will keep your hand and the sandpaper moving at the same speed.

—Jeff McDermott, Phoenix, Ariz.

Spring-loaded dowels

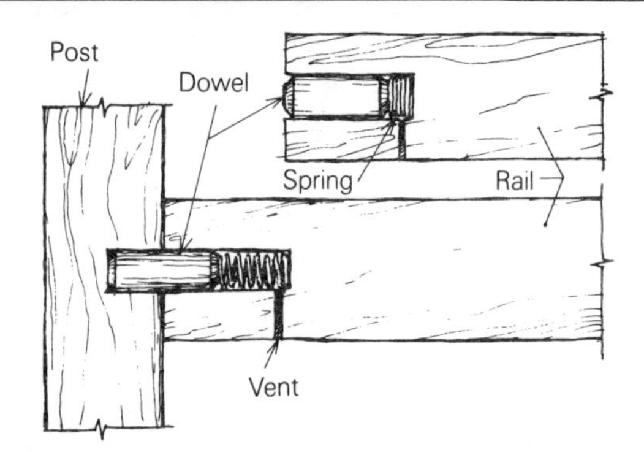

I came up with this simple way to dowel a rail in place between two fixed posts. Drill and dowel one end of the rail as you normally would. On the other end, drill the hole deeper and insert a spring before the dowel. Make the depth of the hole equal to the length of the dowel plus the length of the compressed spring.

Once it's aligned with the hole in the post, the spring-loaded dowel will push its way home. I butter the dowel with hide glue before inserting it in the rail. I also drill a small vent hole in the bottom of the rail to prevent suction problems. Beveling the ends of the dowel allows the parts to go together easily. *—Thomas Ehlers, Austin, Tex.*

Acoustic-tile touchup

Acoustic ceiling tiles are fragile, and they are often dinged during handling and installation. To patch these little scars, I use typewriter correction fluid. It dries fast, and its refrigerator-white color is a good match for the tile.

—T. Marshall Gillum, Orange, Va.

Table-saw moldings

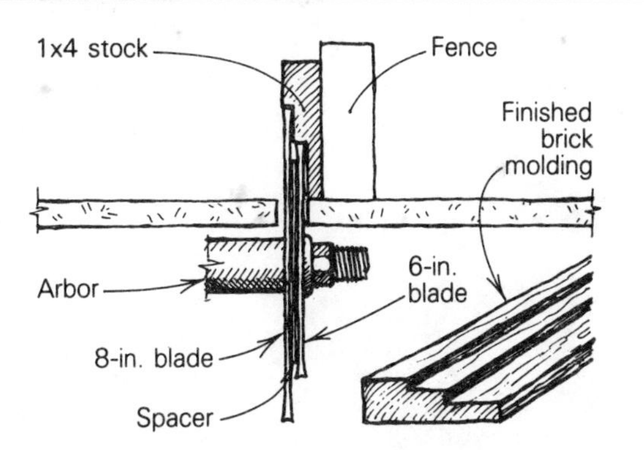

Here's a slick way to make a three-face brick molding in one pass over the table saw. Mount two different blades (I use an 8-in. blade and a 6-in. blade) on the table-saw arbor, with a spacer about 3⁄64 in. thick between the blades to allow for the set in the sawblade teeth. A few pieces of heavy cardstock or building paper ganged together make a decent spacer. Place the fence as shown, and you're ready to roll. You can adjust the face dimensions by raising and lowering the arbor, or by using a thicker spacer or different-diameter blades. ***—Howard Furst, Sedro Woolley, Wash.***

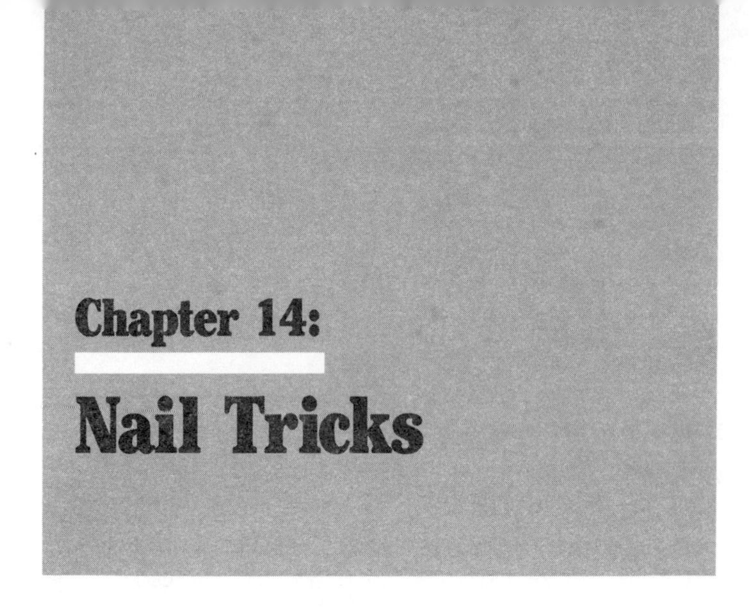

Chapter 14:

Nail Tricks

Tight-spot nailing

The next time you have to sink a nail in a place beyond your hammer's reach, try the technique shown in the drawing. Place the flat end of a 24-in. wrecking bar on the nail head, then hammer the bar shank a few inches from the nail. It works great in tight spots.

—Mike Lyon, Tacoma, Wash.

Starting a nail

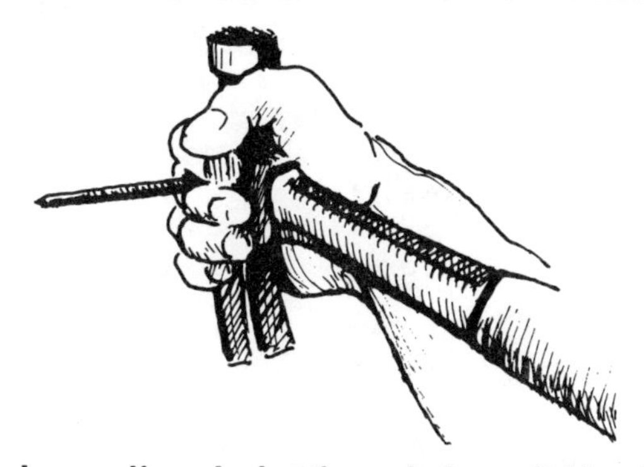

Ever find yourself perched at the end of a scaffold or ladder when you need to start a nail in a piece of framing or siding? If you're like me, chances are you'll be hanging onto the ladder with one hand, leaving you one short to hold the nail. Here's a trick that a veteran carpenter showed me for just such circumstances. Hold the head of the nail against the side of your hammer between the index finger and middle finger as shown in the drawing. Take aim at the spot you want to start your nail and hit it hard enough to make the nail stick. With a little practice you'll have an accurate method of one-handing it. ***—Jim Miller, St. Louis, Mo.***

Bent nail

Bending a box nail on a deck or exterior trim is frustrating even to a veteran builder, but even more vexing is not being able to pull it out.

As an apprentice I learned to deal with this situation neatly and quickly. Using a nailset (2/32 is ideal), dimple the shank of the nail just at the surface of the wood with several sharp blows from your hammer. Then use the claw to bend the nail back and forth without touching the wood. It will break where it was dimpled.

To finish the job, drill a pilot hole right next to the old nail shank and drive a new nail. Its head covers the old scar, and no one is the wiser.

—Claude Foster, Des Moines, Iowa

Cold hands, warm nails

Cold-weather construction generally means a choice between warm fingers and dexterity. Instead of making frequent trips to the site's trash burner to warm your hands, keep a pail of nails warming on the fire. Your nail-fingering hand can be thinly gloved, or not gloved at all, with the nails keeping your fingers warm. This works best when you are using a lot of nails in a short time, but a leather pouch full of warm nails retains its heat longer than you might think.

—***Albert Treadwell, Sandy Hook, Conn.***

Split-free nailing: 3 tips

To minimize the chance of splitting when nailing near the end of a board, make an indentation across the grain of the wood with the head of a box nail. Do this by driving the edge of the nail head into the wood. Then remove the nail and drive it just inside the mark, away from the edge of the wood. The head of the nail after it's driven usually hides the indentation. —***Steve Larson, Santa Cruz, Calif.***

A simple but usually effective trick of the trade I learned from a framer was to blunt the point of the nail before driving it wherever there might be danger of splitting. This can be done by setting the nail point up with the head on a hard surface and striking the point several times with your hammer.

Rather than separating the wood fibers, which may cause splitting, the nail tears the fibers as its blunt end is driven. The nail acts like a punch and makes a cleaner hole. The safest alternative, of course, would be to predrill, but this is often impractical, especially when framing.

—***Robert Nye, Waterford, Conn.***

Most people are vaguely aware that the points of nails are diamond-shaped. However, few people know that split boards can often be eliminated by proper orientation of a nail before driving it. This trick, shown to me by an elderly shipwright, has saved me a lot of predrilling and split moldings.

When driving a nail close to the end of a board, keep the rows of marks left by the nail-forming machine perpendicular to the grain of the wood. This causes the nail point to cut the nail fibers rather than wedging them apart, thus preventing a split from developing. It's easy to feel these ridges between your thumb and forefinger and then roll the nail to its proper orientation. This can be done so rapidly that you won't lose any time, even when framing. Brads and small finish nails inserted in a brad driver can be aligned by eye, since the diamond shape of their points is even more pronounced than on a box or common nail.

—*Steve Harman, Okanogan, Wash.*

Shingle cheater

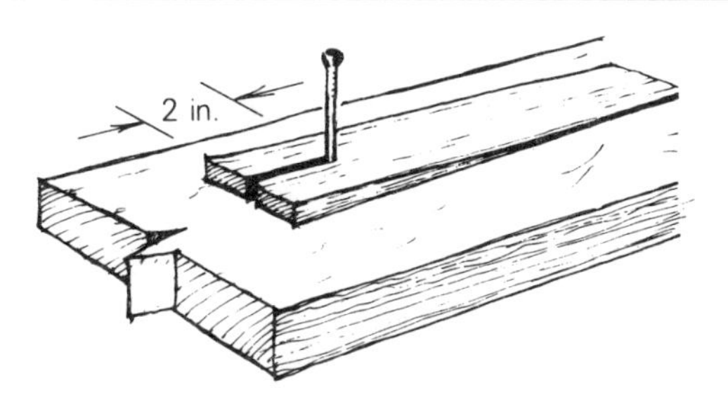

Here's a trick an old carpenter showed me after watching me drive 16d galvanized casing nails, overhead, into 2x6 vertical-grain fir. I had just missed the nailhead, producing a deep hammer track in the board, and I was about to turn the air blue with a few choice phrases when he showed me how to make a cheater. He took a shim shingle and cut a lengthwise slot into the middle of the butt end. Once he had started the nail, he slipped the kerfed shingle around the shank of the nail, as shown, and invited me to hammer away. After that, when I missed the nail, the shingle took the hit, not the trim. —*Jeff Head, Port Angeles, Wash.*

Invisible nailing

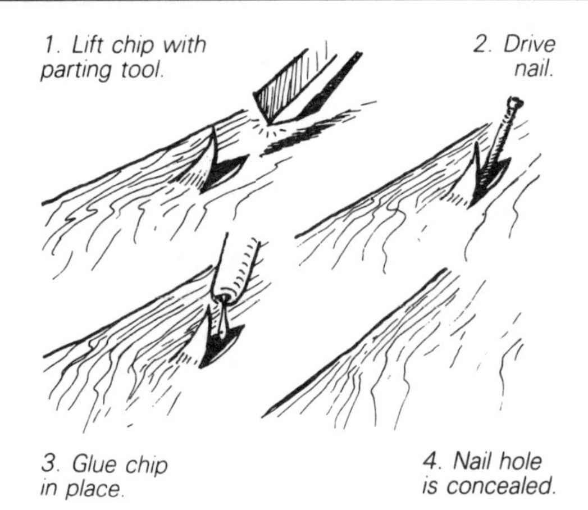

We don't worry about filling nail holes—we eliminate them.

Before driving the nail, we use a gouge or a parting tool to lift a small chip of wood from the surface of the work (1 in the drawing). Be careful not to break off the chip. Now drive the nail into the resulting groove (2), and countersink it. Glue the chip back in place (3), and you've got a concealed nail hole with perfectly matching grain (4).

—Daniel Casciato and Christopher Gecik, Cleveland, Ohio

The pea-shooter

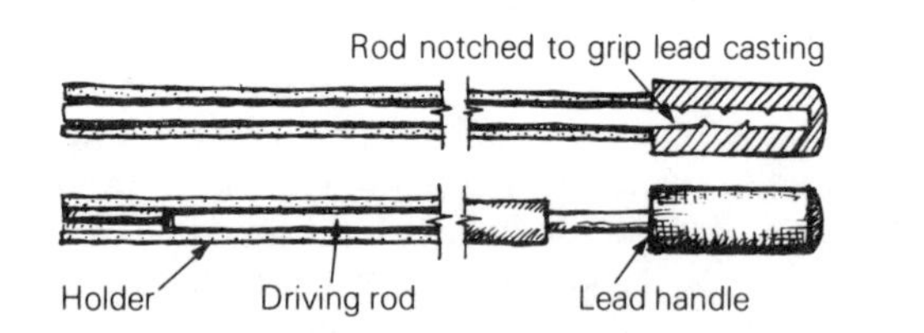

Ever need the help of a 4-in. carpenter to drive the first nail when you're installing cabinets? Here's a useful addition to your tool kit that can help you drive those nails in tight places—the pea-shooter.

The pea-shooter (drawing, above) consists of three inexpensive and easy-to-assemble parts: the handle, the driving rod and the holder. The handle should weigh about 3 lb. and have a rounded end that's comfortable to grip. A lead casting, a bumper hitch ball or a steel bar all make good handles.

The driving rod is about 30 in. long and should be a piece of oil- or water-hardened drill rod. Mild steel rod can be used, but won't last as long. For 16d and 20d nails, use a ⅜-in. rod. For 8d and smaller, use 5⁄16-in rod. One end of the rod is secured by welding to a steel ball or notched and cast in a lead handle. A steel handle can be secured to the rod with setscrews, which also allows the rod to be easily replaced. The end of the rod that strikes the nail should be hardened by heating it over a natural gas flame until it is brownish purple, and then quenching it immediately in oil or water.

The holder is a piece of tubing or pipe with a ½-in. or ⅜-in. inside diameter (depending on the driving rod) with a relatively thick wall. The length should be exactly the same as the length of the driving rod, excluding the handle.

To use the pea-shooter, draw back the driving rod into the holder, insert a nail into the end of the tube, and place the end of the tube where you want a driven nail. A few blows with the driving rod and the nail is home for good.

—Dave Jochner, South San Francisco, Calif.

No-nail zone

An old carpenter once advised me not to use nails in the area 12 in. to 24 in. above the sole (bottom) plate at places such as built-up corners, trimmers or doubled studs. This way, holes for electrical wiring can be drilled at typical outlet heights without running into nails. It sure saves a lot of auger bits. ***—Dick Haward, Nehalem, Ore.***

Carpenter's oil

While bending one 8d nail after another trying to install an oak window skirt without predrilling, I was mercifully shown a valuable trick by an experienced carpenter. He took one of the nails, rubbed it quickly through his hair and then drove it through the oak with no bending at all. He assured me that even bald carpenters can strike oil by using the side of their nose. ***—Charles E. Lord, III, South Portland, Maine***

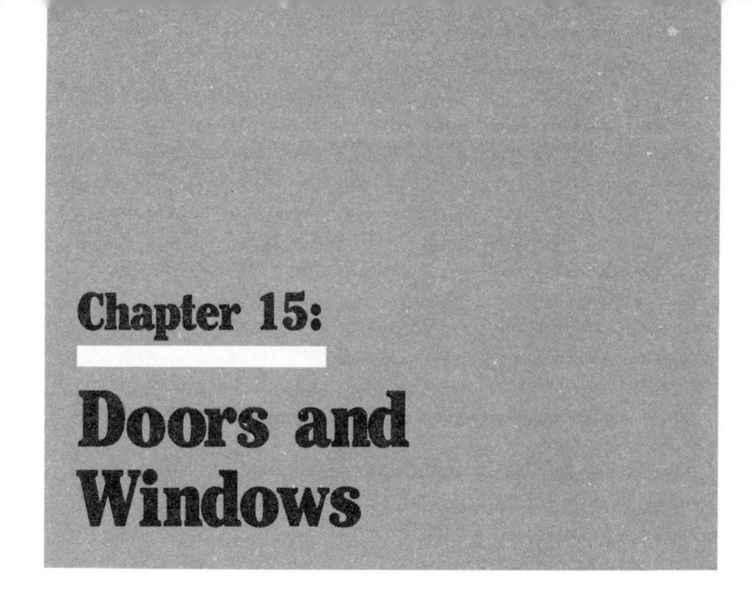

Chapter 15:

Doors and Windows

Holding a level plumb

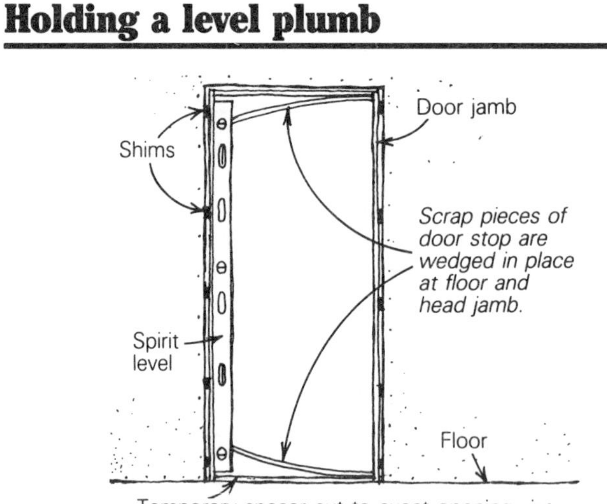

I use a 6½-ft. spirit level to plumb and straighten interior door jambs. In order to leave both hands free to handle the shims and to drive nails, I wedge the level against one of the jambs to hold it in place. This lets me monitor the jamb continually as I make fine adjustments with the shims, rather than having to check and recheck the alignment with a hand-held level.

Once I've assembled all the jambs for the house, I cut a spacer the exact width of the door opening out of 1x6 or jamb offcuts. As shown in the drawing, the spacer rests on the floor between the jambs, maintaining the correct dimension at the bottom of the door opening.

Spacer in place, I temporarily shim the jamb in its rough opening. Then I wedge the level against one of the jambs with scrap pieces of door stop. The spaces above and below the level should be equal. Now I can plumb and straighten the jamb in the usual manner, tapping shims in and out until the jamb is straight up and down and true as an arrow. Then I nail the jamb to the jack studs with a pair of 10d finish nails through each set of shims. After double checking for alignment and making any necessary adjustments, I repeat the procedure on the opposite jamb.

—D. B. Lovingood, Portsmouth, Va.

Entry-door security

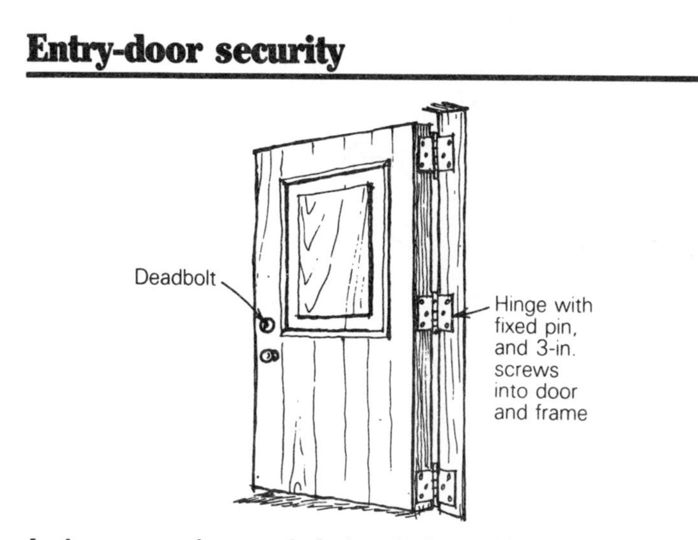

In the course of my work, I often find myself beefing up the security features on entry doors. In addition to adding a deadbolt, I like to add a reinforced deadbolt strike plate. I attach it, and the door hinges, with 3-in. screws.

Some doors are hung with their hinge pins vulnerable to the outside, inviting a break-in. Rather than resort to unsightly pins that protrude from the door jamb, I simply add (or replace) a hinge mid-height on the door with a fixed-pin hinge. Once this hinge is installed with 3-in. screws, the edge of the door is secure from inside or out. A carpenter needing to remove the door simply removes the screws from one leaf, and then has the convenience of two loose pin hinges while planing, or re-hanging the door. Once those screws are back in, the door is there to stay.

—Kurt Lavenson, Berkeley, Calif.

Door-holding jig

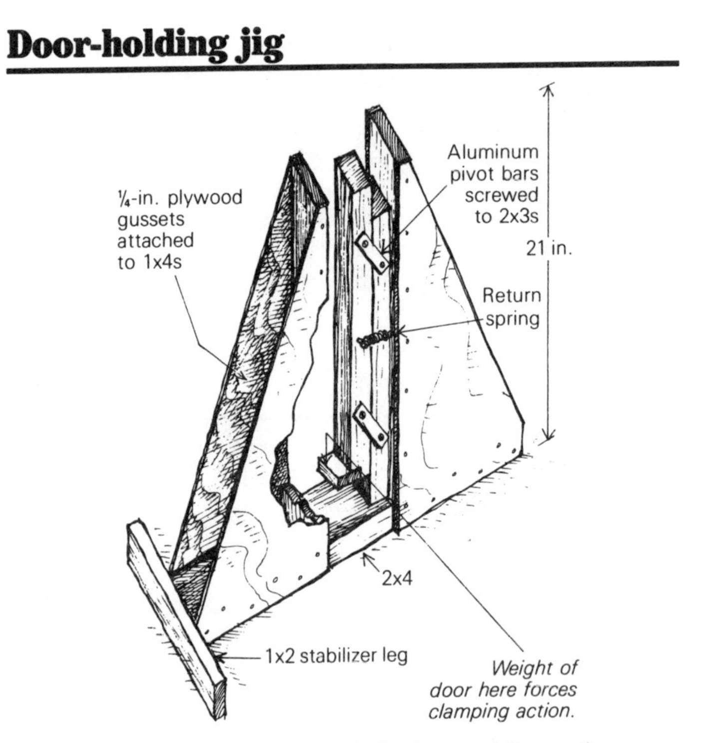

This is a sketch of a jig I use to hold doors while routing them for hinges or planing them to fit. It automatically adjusts to doors of different thicknesses and holds them securely. The weight of the door provides the clamping action, and the return spring brings the clamp back into position.

—Kerry Ludden, Kearney, Neb.

Installing a new threshold

It's important for a door to fit snugly against its new vinyl-gasket threshold, particularly if the system you're using doesn't include a door-bottom sweep. But fitting an old door with one means having to trim the door bottom very accurately.

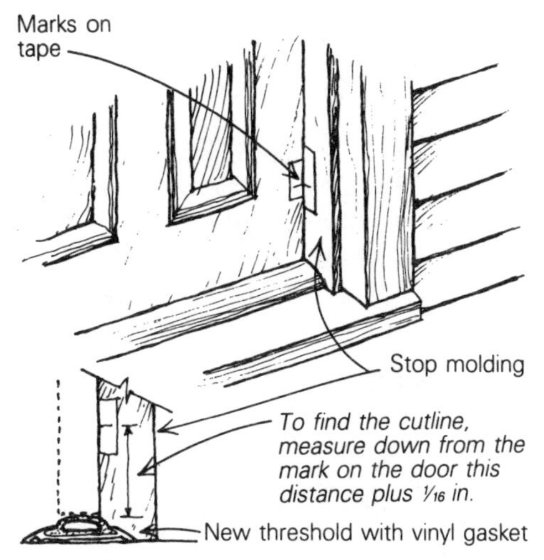

Before removing the door from its hinges to install the threshold, I put a piece of tape down low on the face of the door and another across from it on the face of the stop (or its equivalent on a rabbeted jamb), as shown in the drawing above. Then I make a single pencil line across the two as a registration mark, and I remove the door. Once I've installed the new threshold, I can then measure from the top of the gasket to the pencil mark on the stop and transfer this number to the door by measuring from its registration mark down. I usually add 1/16 in. to get my cutline—this will ensure good compression when the door is closed.

—Rod Goettelmann, Vincentown, N. J.[hrt]

Adjusting doors

Sometimes a new door will hang cockeyed in an old frame, despite the fact that the door and frame are properly sized, square and plumb. This can happen if the depth of the hinge mortises is slightly off, if there's a twist in the jamb, or if there's any variation in the butts themselves.

To remedy this situation, most carpentry books advise shimming the hinges to throw the hinge pins closer to or farther from the jamb. This trial-and-error method takes quite a lot of time. Instead, simply close the door and remove the hinge pins from either the top and middle hinges, or bottom and middle hinges, and temporarily shim the bottom of the door so that the gap is even along both edges of the door. Now take a 6-in. adjustable wrench or smooth-jawed pliers and carefully bend the hinge knuckles until they once again align, and replace the pins. I know that this approach sounds brutal, but it does work. If necessary, you can pull a door as close as 1/32 in. of the jamb without making it hinge-bound. In fact, making minor adjustments with this method will automatically relieve the stiff action of bound hinges without ever having to loosen a screw or cut a cardboard shim. *—Dave Walter, Oakwood, Ill.*

Hinge-mortising jig

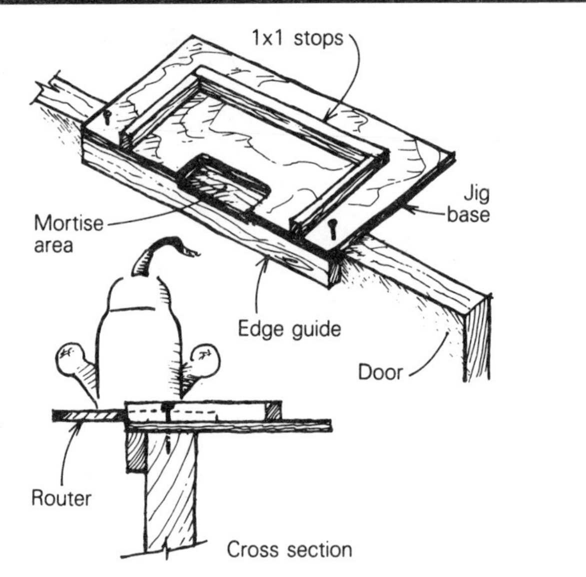

I use a router and a jig to mortise hinges in doors and casement windows. By using a ½-in. diameter hinge-mortising bit and a standard round-cornered hinge, I achieve a consistently clean fit that would be difficult and slow to get by hand.

A piece of ⅜-in. plywood forms the base of the jig. After carefully measuring the size of the hinge and the router baseplate, I nail 1x1 fences to the base to guide the router. As the router is passed inside the bounds of the fences, the bit will cut the hinge shape into the plywood base. Once the mortise is cut out of the base, I lower the bit to adjust the depth of cut in the door.

A 1x2 edge guide nailed and glued to the base aligns the jig on the edge of the work. To keep the jig steady while using the router, I tack 6d nails through the plywood base into the edge of the door or window jamb.

—Steve Larson, Santa Cruz, Calif.

Help for pulled-out hinges

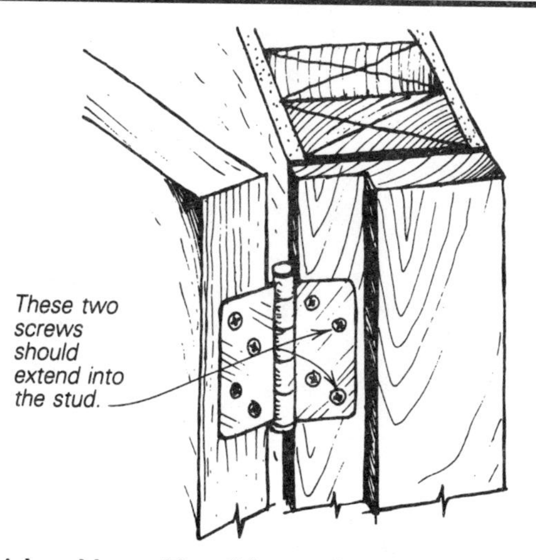

A perennial problem with solid-core doors is that they are so heavy that they pull away from the jamb at the top hinge. To prevent this, I replace the two screws closest to the door stop with two 2½-in. or 3-in. wood screws, as shown below. (Drywall screws aren't a good choice, since they aren't plated to match and will eventually work loose because they aren't tapered.) It is important to get the screws as close as possible to the center of the jamb, or you run the risk of missing the trimmer stud. I hang new doors with the extra-long wood screws as a matter of course, and I find that it saves me a lot of planing on old doors that "don't fit any more." —***Sam Yoder, Cambridge, Mass.***

Renovation metalwork

People who do renovation work or build rustic new houses are often in the market for black hardware—hinges, latches or just ordinary nuts, washers and bolts. But the right sizes may not be available in black, and when they are, they're more expensive than their silvery counterparts. I permanently blacken my hardware with gun blueing. The package directions are easy to follow, and the finish doesn't chip off. Two suppliers are: Birchwood Casey, Eden Prairie, Minn. 55344; and Metal Blueing Products Inc., New Rochelle, N.Y. 10801. —***Gerald R. Jacob, Kaysville, Utah***

Arched window preparation

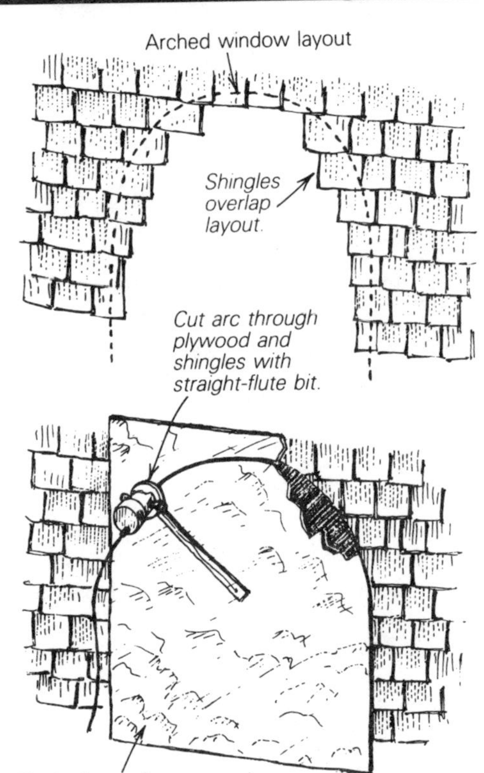

If I'm installing half-round windows or arched windows with a consistent radius, I take care of the wall finish first. Although this method works with all sorts of siding, let's use shingles as the example. Instead of jigsawing each shingle to fit, I let the shingles overlap the rough opening. While doing this I make sure that no nails are driven within a few inches of the rough opening. This ensures an unobstructed space under the shingles for my flashing.

Next I tack a piece of thin plywood (¼-in. plywood will do) over the area to be occupied by the window. This is my flat work surface. Then I attach a trammel to my router, and I screw it to the center point of the arch. Using a straight-flute bit, I make my first router pass deep enough to cut through the plywood. Then I continue with more passes until I've cut through the shingles. If the pivot point is placed correctly, the opening I get will fit my window and its trim.

—David Hornstein, Arlington, Mass.

Window box

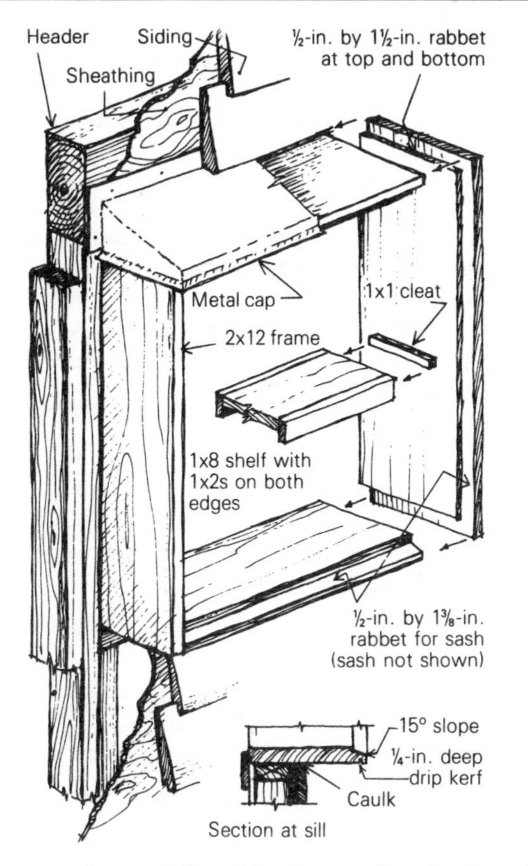

In the course of remodeling older homes, I've developed the window box/greenhouse shown in the drawing above. Its deep recess gives the same comfortable feeling of a window in a thick stone or adobe wall, and makes space for plants and other things.

The heart of the design is a 2x12 frame built with clear, dry lumber and then nailed into the window opening. The frame is easily cut and assembled on the job site with portable electric tools and can be fitted with any combination of casement, awning or fixed sash (new or used). Inside, I trim the edges with casings consistent with other openings in the house. For weather protection I install a sloped metal flashing above the window and caulk along the jambs and sill. To date, all the windows I've built this way have proven to be weathertight. ***—Glen Jarvis, Berkeley, Calif.***

Squeaky hinges

Like anyone else, I used to squirt a drop of oil on squeaky door hinges to quiet them. I would put the oil at the top of the pin, hoping it would creep around the hinge barrel until it soothed the squeaky spot. This works for a while, but inevitably the lubricant wears off, and the squeak comes back.

I've discovered a better way to get the oil to stay where it will do some good, by creating a little oil reservoir at the top of each hinge. I use felt donuts that fit over the hinge pins for this purpose. The donuts go above the top knuckle of the hinge, and they get a squirt of sewing-machine oil or WD-40. If you don't have any felt on hand, try a few twists of cotton string instead.

—W. J. Hutchins, West Hartford, Conn.

Lockset retrofit

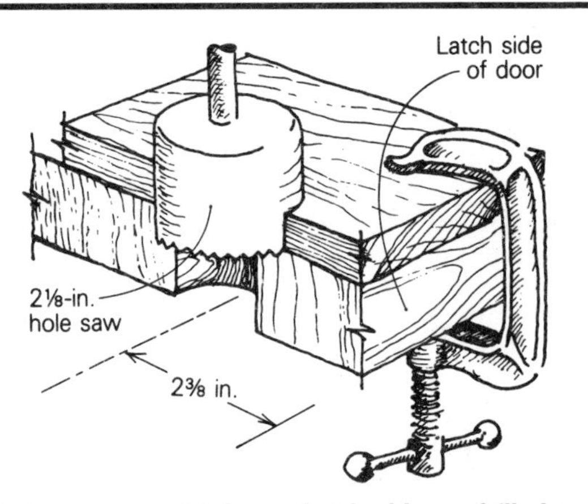

I recently hung some old doors that had been drilled to accept 1¾-in. locksets. These locksets needed replacing. To enlarge the old holes to house the new 2⅛-in. cylinder sets, I had to make a jig that would center my hole saw over the existing hole.

I took a scrap piece of ¾-in. stock, measured 2⅜ in. from the edge to allow for the standard setback, and drilled a hole with my 2⅛-in. hole saw. Then I clamped this guideboard over the existing hole, flush with the edge of the door, as shown above. It acts as a guide for the outside of the hole saw itself, not for the arbor bit, and the holes come out clean and accurate. ***—Mark Messier, Eugene, Ore.***

Site-built leaded windows

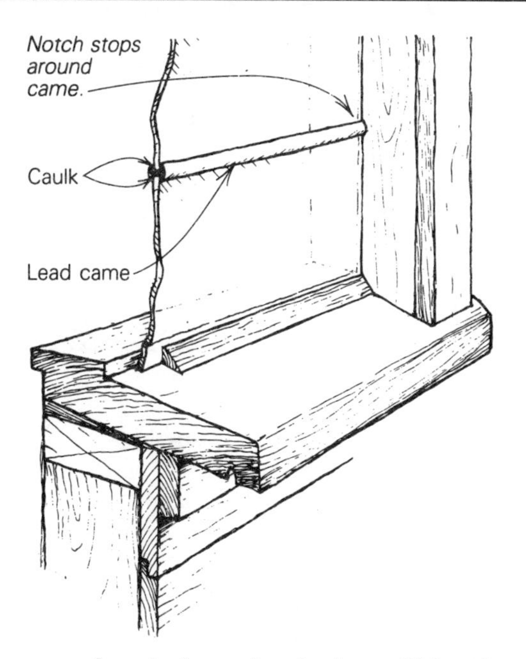

A few years ago, I worked on a low-budget addition that incorporated a lot of recycled materials. One way we were able to save money was by building fixed-lite windows into the bays between studs.

We had been given used windows for the price of hauling them away, but the panes weren't nearly large enough to fill our rough openings. I designed the new windows to take three lites of the recycled glass, separated from each other by short horizontal strips of stained-glass leading (cames). It gives the windows the look of leaded glass, which is consistent with the style of the original house.

For weatherproofing, we laid a bead of caulk in the grooves of the cames before the glass was slipped into place. Both interior and exterior stops are notched around the cames. ***—Robin Mark Freeman, Berkeley, Calif.***

Removing window glass

Extracting the glass from discarded windows for re-use is a sensible idea; many times the glass can be cut to fill a gap somewhere around the house. The only real drawback to window recycling is getting the glass out of its old frame in one piece. Rock-hard putty can sometimes make it nearly impossible. Here's how to do it:

Lay your window down on a flat surface and paint the old putty with lacquer thinner. Then wait five or ten minutes, and go after it with a scraper. The thinner penetrates underneath the putty, allowing it to be lifted away in large chunks. I've never seen putty that wouldn't succumb to this treatment. *—Ron Davis, Novato, Calif.*

Door butt gauge

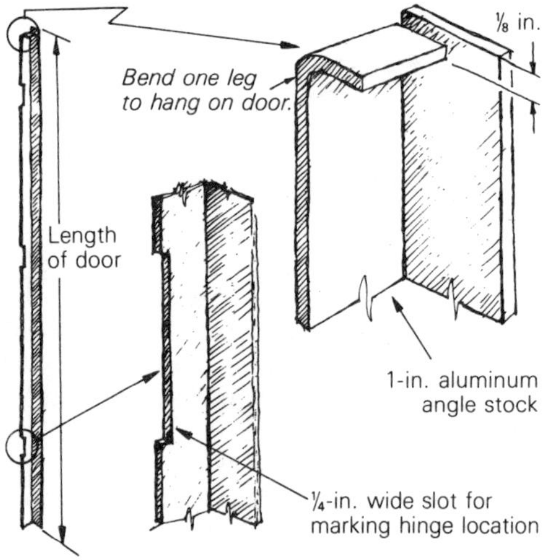

For accuracy and speed when hanging doors from scratch, I use a butt gauge (drawing, above) that I made out of 1-in. aluminum angle stock. Bending one leg of the angle over 90° just ⅛ in. down from the top, and cutting slots the same size as the hinges where they are to be positioned on the door, complete the gauge.

To mark the door, just hook the bend over the door edge and scribe the hinge cutouts with a utility knife. For the hinge jamb, hold the unbent face of the top of the gauge to the top of the jamb and mark the butt spacing. When you are not hanging doors, the gauge and a couple of clamps serve as a handy straightedge for cutting and scribing lines on almost anything. *—John Toly, Richmond, Calif.*

Bifold bumper

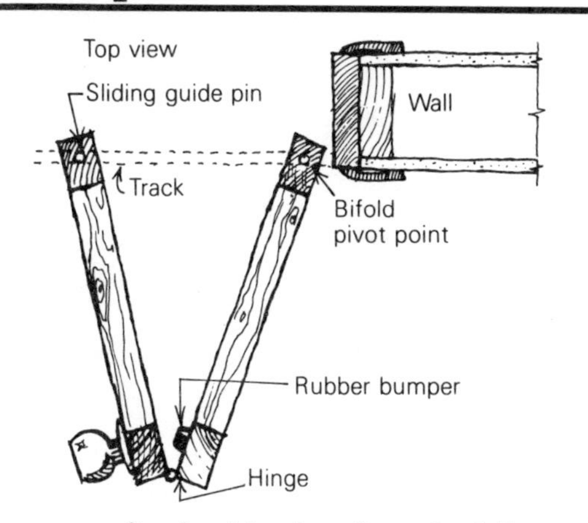

Bifold doors are often hard to close from the fully open position. Some need a pull from the inside back of the lead door to close them. The cure I've come up with is a simple ⅝-in. diameter rubber bumper placed near the bottom hinge on the pivot door. The bumper prevents the doors from coming too close to each other, allowing their normal operation. *—David Graper, Painesville, Ohio*

Screw-clamp door buck

Sometimes we forget that a tool can be used to advantage for other than its original purpose—the ubiquitous screw clamp, for instance. I needed to plane the edge of a small door recently and found a pair of these clamps to be handy supports. By clamping one to each end of the door, as shown above, I was able to brace the door while I planed its edge. *—L. Fredrick, Aspen, Colo.*

Pocket-door precautions

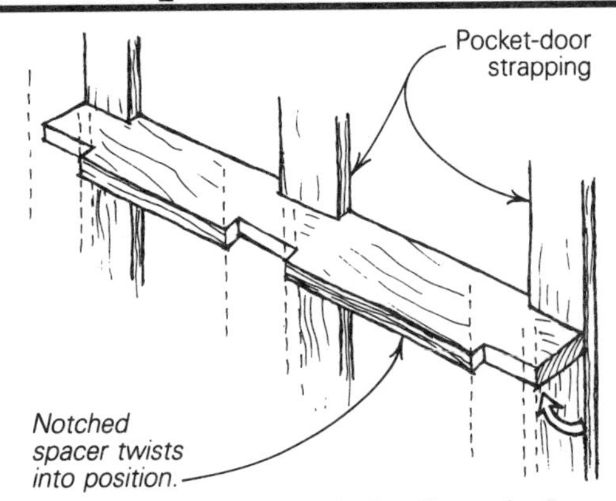

Have you ever had problems with the flimsy 1x flat studs (strapping) that come in a pocket-door kit? Sometimes a sheetrocker will knock one out of alignment, and the door will rub against it. Here's a temporary preventive measure that will hold the strapping in position.

Trim two 1x4s to the appropriate length and cut notches in their sides to accommodate the wall-support straps. These will be used as spacers, so they should be snug but not too tight. Slip these spacers into the pocket on edge, and twist them into position. One in the middle and one at the bottom are usually enough to stiffen the frame while drywall work takes place. When the walls are done, twist the spacers out and put in your door.

—Marvin Havens, Greenville, S. C.

Sill seal

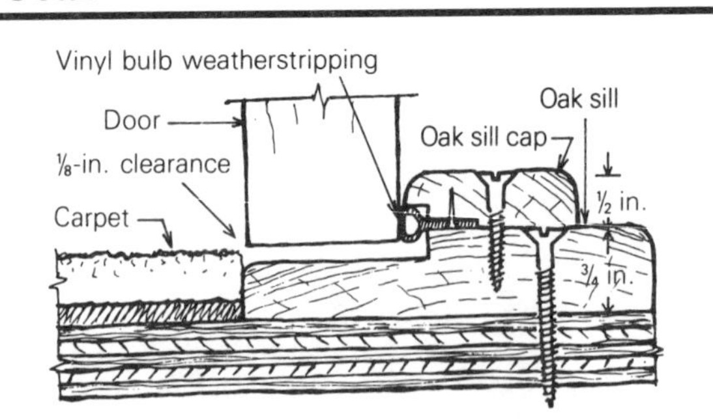

I came up with a solution for weatherstripping a door bottom where a thick rug prevents a spring-loaded bottom sweep from retracting. I used a length of vinyl bulb weatherstripping, intended for a jamb installation, between a two-piece oak sill that I made up in the shop. I tacked the vinyl in place with brads, and then fastened the cap piece to the oak sill with brass screws. This way the vinyl can be easily replaced when necessary. However, there's no foot traffic on the vinyl itself, and it's above the grit line, so it should last quite a while. *—Felix Marti, Monroe, Ore.*

Adjustable counterweights

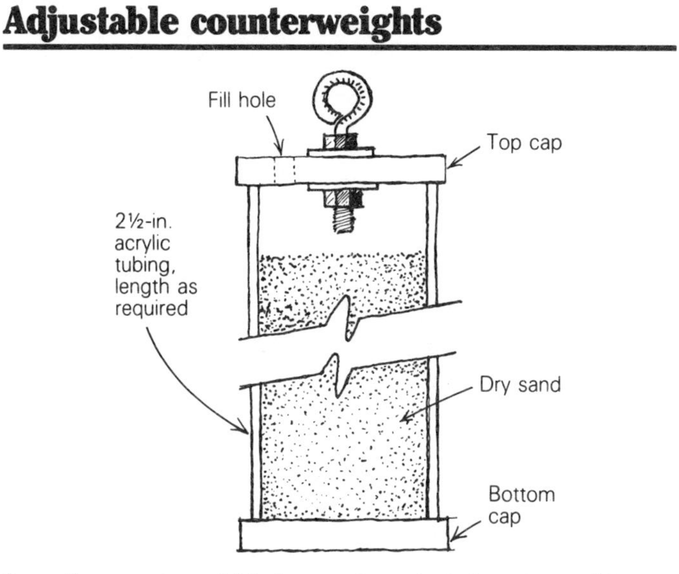

I use the counterweight shown above to balance movable insulated shutters. The beauty of this weight is that I can adjust it while it's in position, rather than running in and out to the truck to file away at a piece of iron bar stock.

The basic container is a piece of 2½-in. dia. clear acrylic tubing; tubes between 1 ft. and 2 ft. long will do for most shutters. I glue acrylic caps to each end of the tube; the upper cap incorporates an eyebolt with a pair of nuts and washers.

To load the counterweight, attach it to the shutter's draw rope and make sure the shutter is in its open position. Insert a funnel into the fill hole in the cap, and slowly add dry sand. When the shutter starts to move, stop adding sand, and pull on the rope to bring the shutter to its closed position. It should stay put. If not, adjust the counterweight accordingly, and cap the filler hole with a dab of silicone caulk. The finished weights can be dressed up with a tube-sock covering, or you can use colored sand or marbles for ballast.

—Bryan Curry, Attica, N. Y.

Radiused window casings

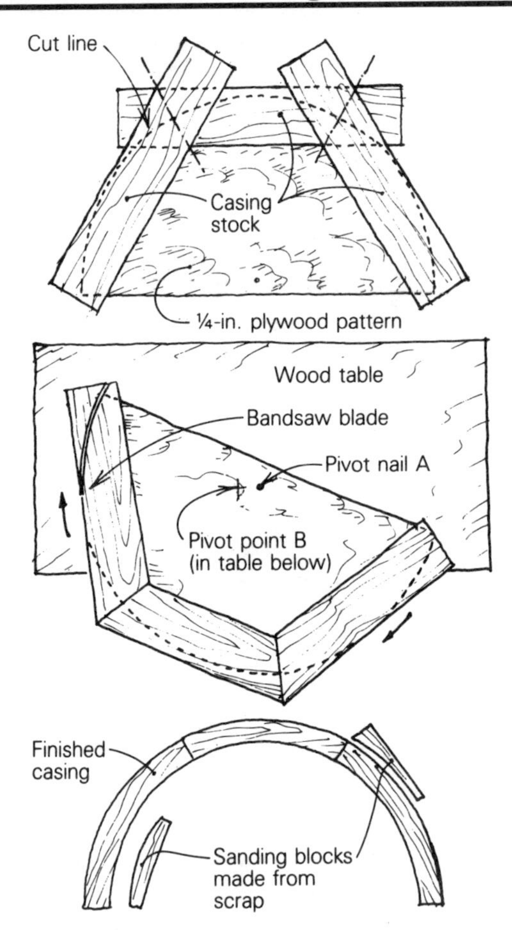

Recently, while building a half-circle window, I needed a way to make the casing and interior stops. Instead of transferring the curves from templates onto the stock and then joining these pieces, I used a bandsaw fitted with a wood table, and a circle-cutting jig.

In this project, the interior face of the window jamb had a radius of 21½ in. Since the exterior casing was to act as the outside stop, the inside radius of the casing needed to be 21¼ in., allowing ¼ in. to overlap the window jamb. The casing was to be 2 in. wide, so the outside of the casing would have a radius of 23¼ in. I cut a half-circle with a radius of 23¼ in. from a piece of ¼-in. plywood by drilling a hole near its center and fitting it over a pivot nail (A in the drawing) that was attached to the wood bandsaw table

23¼ in. from the blade. On this pattern I placed three pieces of stock, arranged to avoid short grain, and marked the necessary cuts. I tacked them to the plywood and ran the pattern with the three pieces attached to it through the bandsaw, again from pivot A. This gave me my outside radius. I then repositioned the hole in the circle-cutting jig to pivot point B, a radius of 21¼ in., and ran the pattern through again. This gave me my 2-in. wide casing with the joints already fitted. I saved the waste from both cuts, and used them as sanding blocks for the inside and outside edges of the casing. ***—Brian K. Shaw, Johnson, Vt.***

Frictionless hinge

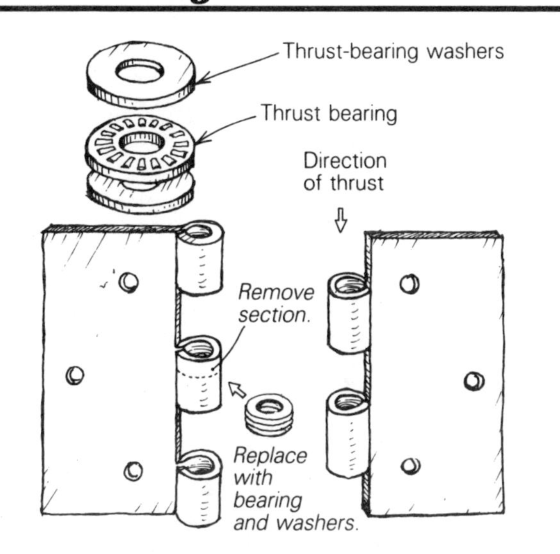

For me, the most frustrating part of making and installing a heavy, handmade entry door has been finding the right hinges. Even with a good set of heavy butts, the hinge barrels begin to grind down after a while. Soon, the door sags so much that the bottom rail drags on the threshold. A solution to this problem became obvious when I discovered thrust bearings at a bearing-supply outlet recently. Thrust bearings (drawing, above) are simply a set of needle bearings captured in a circular metal collar. They come in a variety of sizes. These bearings are used in conjunction with a pair of thrust-bearing washers, which are machined to close tolerances.

To fix the door, I used a hacksaw to cut a section out of

the appropriate barrel of each hinge leaf, replaced the section with the thrust bearing and the washers, then reinforced the hinge pins, as shown. My hinges show virtually no wear after six years of use. —***Clint Lewis, Woodland Park, Colo.***

Center-pivot doors

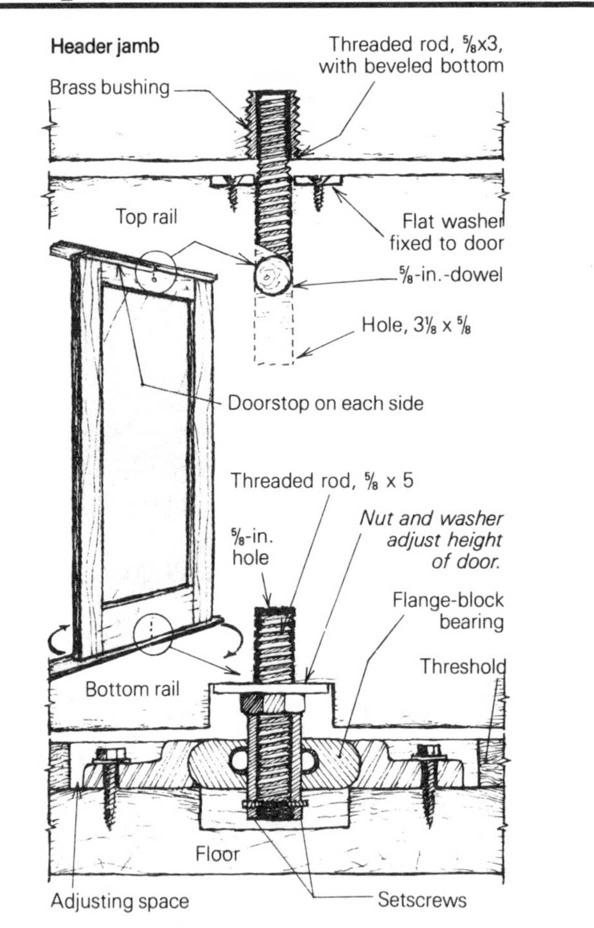

I've always admired Frank Lloyd Wright's use of center-pivot doors arranged in a row. When open, they show only their edges, allowing the wall to disappear.

Recently I wanted to use three such doors to divide a living room from an attached solar greenhouse, so I hunted around for the proper hardware. Ready-made center-pivot hinges were available, but they were designed for heavy commercial use and cost a hefty $200 apiece. Discussion with some friends led me to an industrial equipment supplier, where I found something called two-hole, self-aligning flange-block bearings, at only $12 apiece. With these bearings, another $10 worth of parts from the hardware store and a liberal amount of head-scratching, the doors were swinging smoothly on their hinges.

I cut an oversize mortise for each bearing block into the threshold and floor to allow a measure of adjustability, as shown in the drawing. I used threaded rod for the top and bottom pins, and locked each bottom pin into its bearings with setscrews. A nut threaded onto the pin rests on top of the bearing's inner race sleeve and carries the door load with the help of a large flat washer.

The self-aligning feature of these bearings allows the shaft to swivel, making installation of the door easier by letting you tilt the bottom pin out (this can also be done by leaving the mounting screws out and tilting the whole bearing until the door is in place). The screw holes are mounted parallel to the door so they are exposed when the door is open.

During installation the top pin drops into a hole drilled in the top rail. When the pin is in place, it is retained by a dowel. Once the door was in position, I inserted a screwdriver into the dowel hole, levered the pin into its jamb-mounted brass bushing (an outside-thread plumbing nipple) and then drove the dowel home.

—Robin Mark Freeman, Berkeley, Calif.

Sealing around windows

When I was remodeling a house built in the 1920s, I made the decision to replace all of the leaky, double-hung sash with new windows. My new aluminum windows came with 1½-in. wide flanges that, in new construction, are covered by the siding. In remodeling, however, the flange is nailed on top of, or next to, the old siding, and it has to be covered by trim. However, the trim isn't enough to solve the problem of leakage around the flange. To make matters worse, the building that we were working on had deeply relieved triple-lap siding. A piece of trim nailed to the old siding simply sat atop the high points, leaving enormous gaps that invited water damage.

One possible solution would have been to stuff caulking into all the gaps, but my carpenter had a better idea. He attached 2-in. wide strips of Peel-'n-Seal (Hardcast Inc.,

8242 Moberly Lane, Dallas, Tex. 75227) all around the windows, covering both the flange and the old siding. Peel-'n-Seal, which comes in rolls that range in width from 2 in. to 12 in., is a strip of aluminum covered by a ⅛-in. thick layer of asphalt adhesive. It can be easily tooled to conform to any surface, and it will seal wide gaps. We finished our installation with wood trim, entirely disguising our solution.

—J. A. DeCecco, Santa Cruz, Calif.

Snug screens

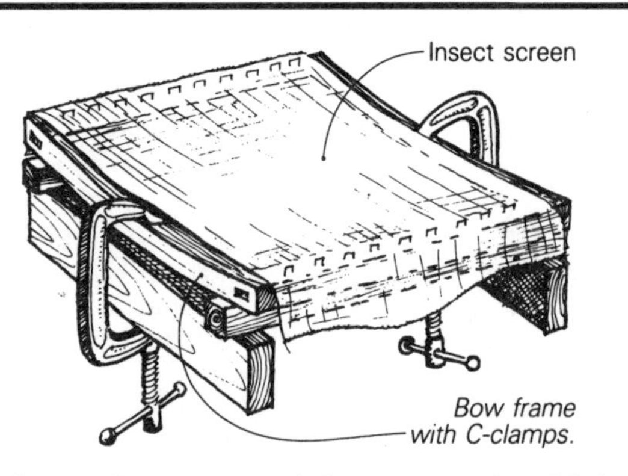

I've found a good way to stretch insect screening tightly and evenly across wooden framing members for screened doors, windows and porches. First I place the empty frame on top of a pair of 2x2 supports. The supports in turn rest on a rigid platform. I use two C-clamps to bow the frame downward, and I staple the screen to the unbowed framing members on the ends. Once the clamps are removed, the frame provides its own tension, tightening the screen. Then I staple the screening to the side rails, and trim its edges with wooden moldings. ***—M. Felix Marti, Monroe, Ore.***

Trimming pocket doors

If you have to trim more than 1 in. from a pocket door (to clear a carpet, for example), do it in place. First, mark your cut with a knife and place your circular saw on the floor, blade low and horizontal, as shown in the drawing. My large saw clears the floor by 1½ in. and my little one one by 1¼ in., which allows plenty of room for the carpet. Position

the saw with its foot flush with the face of the door and open the blade guard. Turn on the saw and, holding it firmly, have someone slowly close the door. This will cut most of the door bottom; the last bit can be trimmed with a Sawzall, a jigsaw or a key-hole saw.

—W. Peine, Eureka Springs, Ark.

Door-trimming jig

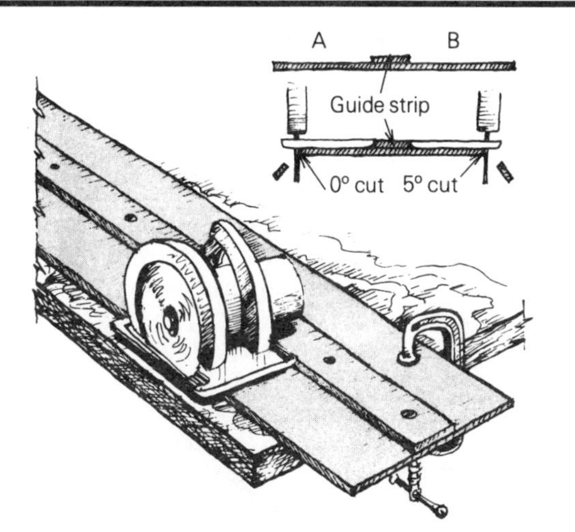

The hinge side of a door must the trimmed to width at 90°. The latch side is usually trimmed at 5° off vertical for a beveled edge. Making these long cuts straight and smooth ia easy with this two-sided jig.

First, buy a fine-toothed blade for your circular saw. I prefer a carbide-tipped plywood blade. Mount it on the saw arbor, and then use a combination square to make sure that the saw's base is 90° to the blade.

Build the jig with ¼-in. Masonite, 8 ft. long and about 12

in. wide. Accurately place a straight Masonite guide strip, 2 in. wide, down the middle, gluing and tacking it firmly in position. Leave dimensions A and B large at first. When the glue is dry, trim the sides with your saw—one side with the blade set at 0°, the other side with the blade set at 5°. Mark the respective sides clearly. When the jig is clamped to the work, the appropriate edge will register the line of cut. For smooth operation, apply furniture wax or car wax to the surfaces on which the saw's baseplate slides.

For crossgrain cuts, first run a knife along the cutline to sever the surface fibers of the wood. This will eliminate splintering, which can be a problem with veneered doors.

—Philip Zimmerman, Berkeley, Calif.

Installing fixed windows

As builders of passive solar homes in a severe climate, we combine heavily insulated concrete and stud walls with fixed glass windows on the south side for direct gain.

Large areas of insulated glass can be installed quite economically if the units are standard sizes and fixed directly to the building's framework. We use tempered glass made for patio doors; they are available at reasonable cost and resist almost any form of abuse. If a unit must be replaced, it can be done easily using this form of installation.

Frame the rough opening with a ½-in. gap at the top and sides to allow for expansion and framing irregularities. Cuprinol or copper naphthalate preservative on the interior trim will reduce the water damage from condensation, which is inevitable. A few coats of marine varnish will extend the life of the wooden sill.

Butyl tape greatly simplifies glazing. We use Butyl 440 tape from Tremco; it is a uniform bead of butyl caulk with a paper backing that prevents it from sticking to itself. It is packaged in coils and can be purchased at many hardware or paint stores. Automobile glass installers also use it. Leaving the protective paper layer on allows you to install the tape against the interior stops, check the window for fit, remove the paper and instantly secure the window. We place neoprene setting blocks under the window as it is installed to absorb expansion and contraction of the glass and frame. The blocks are site cut to the thickness of the window unit from ¼-in. thick neoprene and are normally 1¼-in. long—usually two per panel. Then we use Tremco mono caulk to

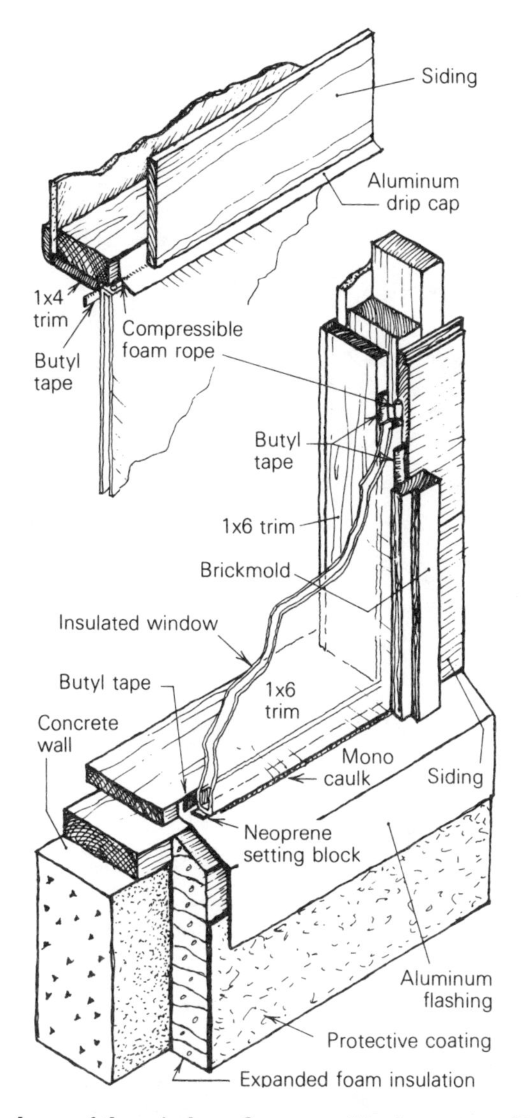

seal the base of the window. Compressible foam rope ¾ in. wide fills the spaces at the sides and top of the window and allows for expansion.

The aluminum drip cap at the top of the window overlaps the brickmold trim on both sides, minimizing exterior trim and further reducing weathering problems. As soon as the sun comes out, any drift of snow or ice at the bottom of the window slides off the aluminum flashing (preferably a dark color). —***Tom Ashman, Wiarton, Ont.***

Chapter 16: Built-ins

Cabinet-door hanging jigs

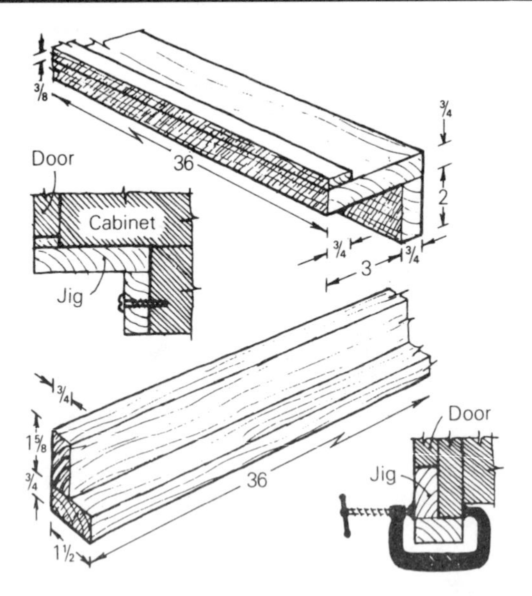

Hanging cabinet doors is easy for most three-handed craftsmen: One hand keeps the door plumb and flush to the cabinet frame, one keeps the bottom of the door at the right height above the bottom rail (so that there is no binding), and one screws the hinges in. For those of us with two hands, the jigs I use help; both work well for lipped or overlay doors.

I rarely hang cabinet doors before the units are installed on the site, although I do fasten the hinges to the door. It's common practice to pre-hang doors in the shop, remove them for transit, and then rescrew the hinges. I see little gain in pre-hanging, for there are always adjustments to be made after the stress of moving racks the frame slightly. I prefer to do the job once.

The first jig shown supports the doors on the base cabinet. The dimensions given suit my purposes, but can be changed to suit your own needs. Using the jig is simple. Slide it into the kickspace and push it tight against the underside of the cabinet edge. To hold the jig in place, screw its back edge to the plinth (the screw holes will be covered by a kick strip) or wedge the jig tight with scraps built up from the floor. With the bottoms of your doors resting on this level jig, you need worry only about the doors' left-right positioning; all their heights will be the same.

The jig for the upper cabinets can be held in place with C-clamps. The jig's vertical dimension is suited to the 2-in. thickness that I favor for the bottom rails of upper cabinets.

—J. E. Gier, Mesa, Ariz.

Matching marble

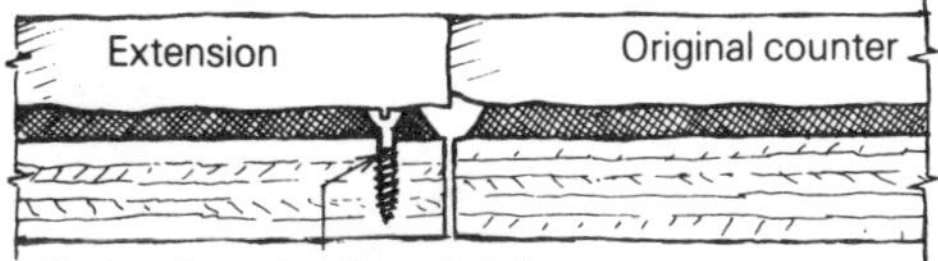

Flathead screw adjusts height.

Our firm does a lot of kitchen remodeling, and recently we had to extend an existing marble countertop a few feet. Locating a similar piece of marble was hard enough, but the real test came during installation; any unevenness in the glossy surface would be quite noticeable.

We got a precise fit using a flathead screw under each corner of the new piece. The screws were easily adjusted to the right elevation, and they kept the heavy marble from sinking into the slowly curing mastic.

—Max Durney, Petaluma, Calif.

Another cabinet-door jig

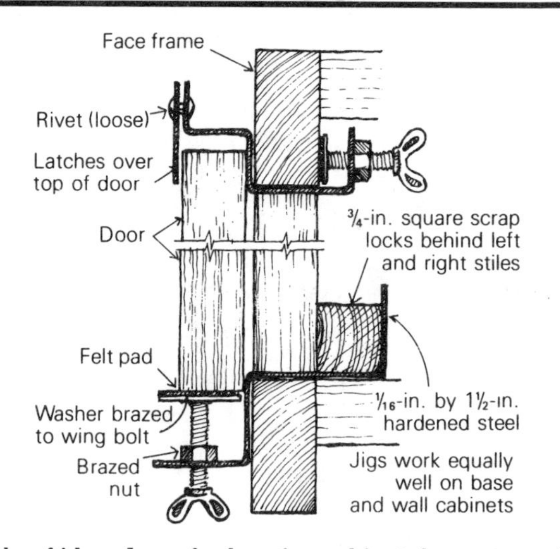

A couple of ideas I use for hanging cabinet doors: two of the bottom jigs level the door; the top jig holds the door up, freeing both hands. With lipped doors I use one bottom jig in the center of the door. The jigs are 1⁄16-in. by 1½-in. hardened steel bent to specifications at a machine shop.

—***Ben James, Jacksonville, Fla.***

Door-pull jig

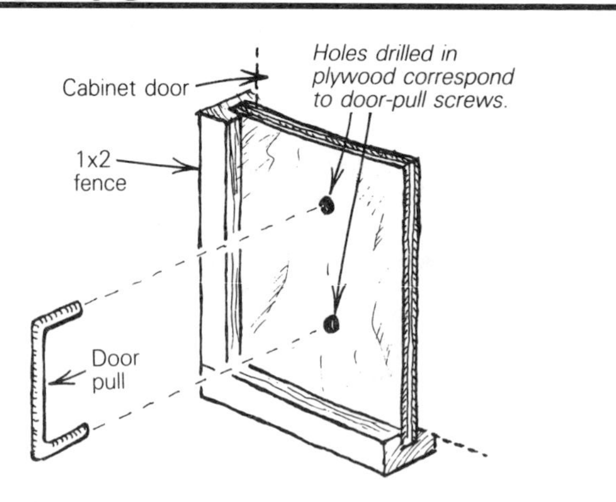

Whenever I need to install pulls on cabinet doors, I begin the job by making a jig to locate the screw holes. The jig (drawing, facing page) is a piece of thin plywood (¼ in. to ½ in. works fine) bordered on two sides by 1x2 fences that meet at 90°. Grooves in the 1x2s accept the plywood insert, creating a fence on both sides of the jig. Once I have decided where I want the pulls to be in relation to the corner of the cabinet doors, I drill corresponding holes in the plywood insert, as shown.

To use the jig, simply snug the fences against the corner of the cabinet door where you want to install the pull, and drill your holes using those in the plywood as a guide. For the adjacent door, flop the jig and you're ready to drill.

—Andrew George, Richmond, Va.

Cabinet wall scribe

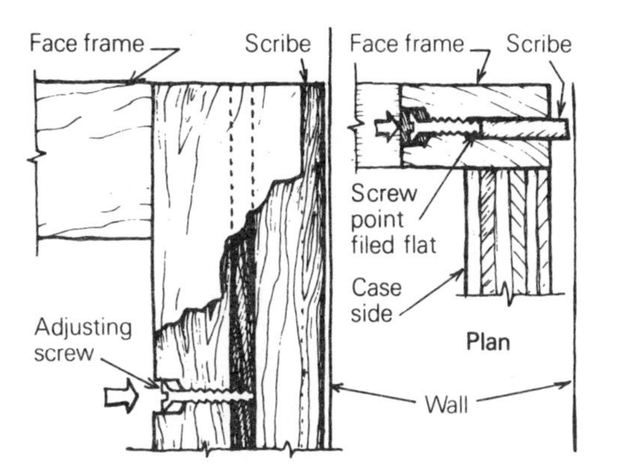

If you don't mind a slight shadow line where the face frame of your cabinet butts a wall, try this system instead of the traditional method of scribing a pencil line on the face frame and cutting it to fit the contour of the wall.

I use a permanent wooden scribe housed in a slot in the outside edge of the face frame to fit my cabinets to the wall. Since this wooden scribe is located in the center of the frame, it produces a slight indentation (or reveal) from the face of the frame. In plan, this system looks like half a spline joint. I cut the slot the full height of the frame ¾ in. deep and ⅛ in. wide. I cut the wooden scribe the same dimensions for a friction fit. Screw holes and screws are then added from the inside edge of the frame on center with the slot.

To use the scribe, place the cabinet ¼ in. from its wall, and then drive the screws. This forces the scribe out of its slot and against the wall. Grind the ends of the screws flat so they don't enter the scribe.

This system copes with out-of-square conditions, and small contours as well, because of the flexibility of the scribe. Spring steel clips can be used instead of screws to hold the scribe to the wall. *—Michael Lynch, San Francisco, Calif.*

Cabinet cleats

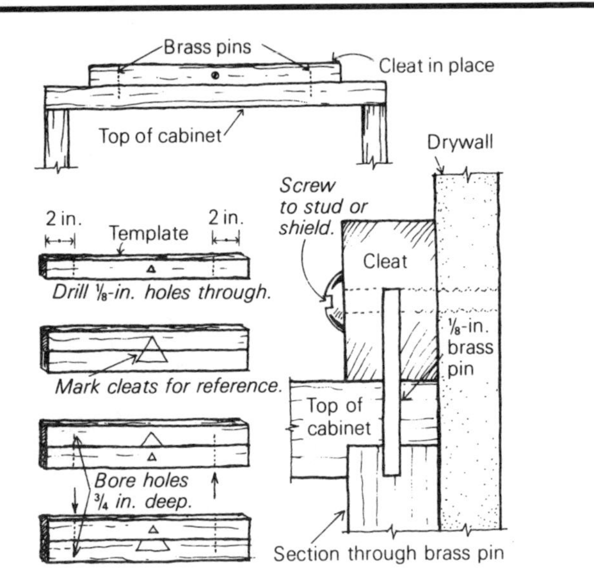

This easy and attractive way to hang cabinets involves a wooden cleat above and below the cabinet against the wall. These cleats are picked to complement the grain of the cabinet and shaped accordingly. They need not be thick or wide. I make three pieces, usually about ½ in. to ¾ in. by 1¼ in. and as long as looks good. One piece is the template. Holes ⅛ in. in diameter are drilled through the template about 2 in. in from the ends to accommodate ⅛-in. brass rods.

Next I mark the top and bottom pieces and the template with a triangle for reference, then bore the two cleats about ¾ in. deep. Next the template is clamped to the cabinet top and bottom and holes bored there. Make sure the template is in the exact position you want your cleats.

At this point cut your pins and try on the cleats. They should fit like a glove. For most of my small cabinets I use #10 or #12 round-head brass screws in the center of each cleat. I like this technique for drywall mounting because you can set the cabinet where you like and mark the screw holes. If there is no stud (is there ever?), you can use those little plastic shields that work incredibly well and come in all sizes. On large cabinets, more pins and screws will make it easier for you to sleep at night.

These cleats can be clean and straight and functional, or fun and frivolous. They can be shaped and sculpted to look like a natural extension of the cabinet top and bottom or can have other roles such as little shelves. Experiment.

—Alan Miller, Brush, Colo.

Multi-purpose boxes

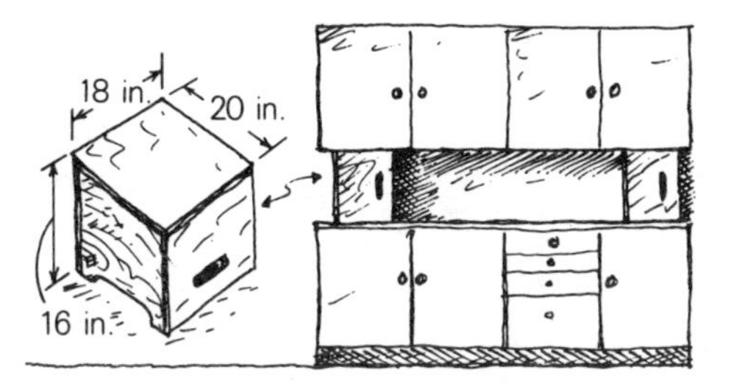

I recently installed a lot of cabinets with the help of some simple plywood boxes. As shown in the drawing above, I made the boxes out of scrap pieces of ¾-in. plywood. Their dimensions correspond to the typical counter-to-overhead-cabinet heights.

While the boxes make it easy for one person to position overhead cabinets, they also have some other uses. When spanned with a plank they make a sturdy platform for working on door casings and crown moldings. Their cutout handles turn them into convenient containers for carrying tools and materials. Finally, they make handy stools at lunchtime.

—Dan Jensen, Portland, Ore.

Gluing plastic laminate

While covering some site-built cabinets with plastic laminate, I tried the often recommended method of laying down brown paper between the laminate and the substrate to adjust the alignment before the contact cement on the two surfaces touch. I had difficulty removing the paper while keeping the laminate in place, and it seemed like a waste to throw the paper away after a couple of uses.

Instead of brown paper, I began using lengths of an old metal tape from a tape measure. Now I place these pieces of blade directly on the glue surface, position the laminate, and then grip the blades with a pair of pliers to slip them out one by one without any problems. The blades are reusable.

—Walter J. Hutchins, West Hartford, Conn.

Cabinet jacks

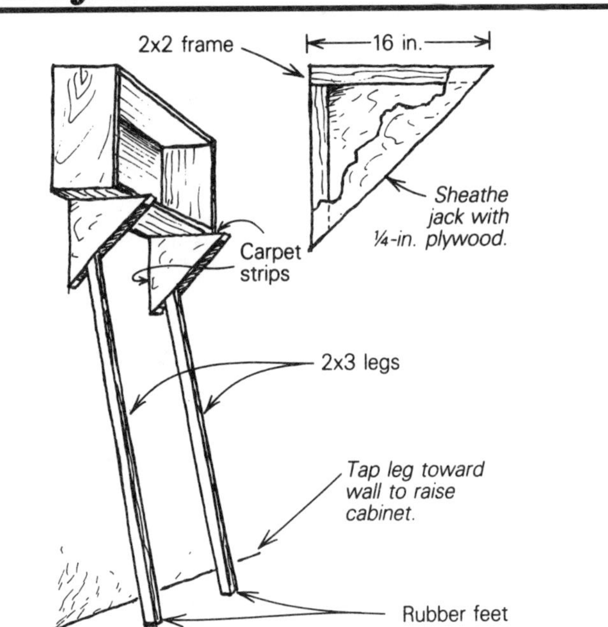

The drawing above shows the jacks I use to install upper kitchen cabinets by myself. The jacks have a 2x2 frame covered with ¼-in. plywood, and they're braced with 2x3 legs. I line the backs and tops of the jacks with strips of carpet to keep them from scratching the walls, and I've got rubber feet on the legs to keep them from slipping as I adjust the height of the cabinets.

To use the jacks, I mark a level line on the wall to show the bottom of the cabinet. Then I place the jacks about 6 in. inside each end of the cabinet, and I set the cabinet on the jacks. Moving the legs in or out adjusts the height of the cabinet. When I've got it right, I fasten the cabinet to the wall through predrilled holes in the cabinet hang-rail.

—Darryl B. Weiser, Dahlonega, Ga.

Hidden shelf connections

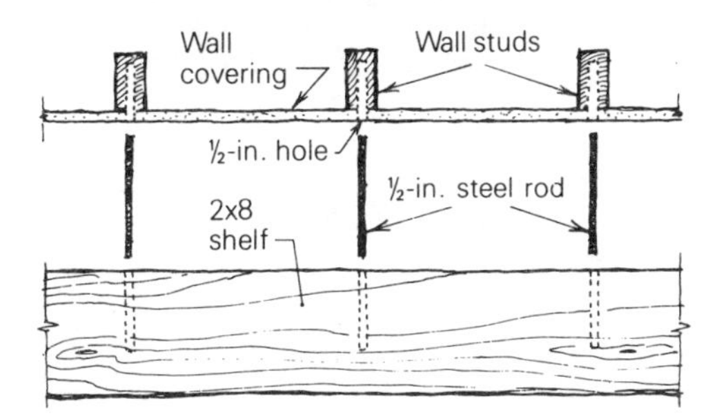

I was working on a game room where I had a 9-ft. section of wall on which I wanted to place shelving. I was doing the room in Early American so I wanted to use 2x8 material to achieve a massive look. I didn't want any visible bracing that would interrupt clean lines. The solution: I drilled ½-in. holes in the wall studs to a depth of 3 in. plus the thickness of the wall covering, on the centerline of each shelf and the center of each stud. I then drilled ⅝-in. holes 5 in. into the shelves, to correspond with the holes in the wall. I drove 8-in. by ½-in. dia. steel rods into the wall studs and then slid the shelves onto the protruding steel rods. The holes are larger than the rods so the shelf can be removed for painting. The resulting shelving is strong enough to sit on, yet still has clean, unbroken lines.

—Stan Morgan, Mapleton Depot, Pa.

Kitchen flashing

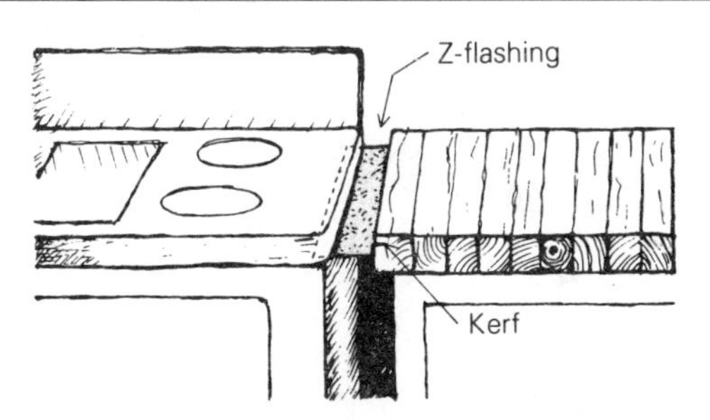

I put a chopping-block counter next to my stove and left about an inch between the two so the block wouldn't get scorched. The extra counter surface was handy, but food and utensils kept falling into the space between the two.

I solved the problem with a 2-ft. length of Z-flashing

placed under the stove lip and into a kerf cut in the chopping block, as shown in the drawing. No more lost butter knives or green peas rolling under the stove.

—Bruce McGaw, Albany, Calif.

Two more cabinet jacks

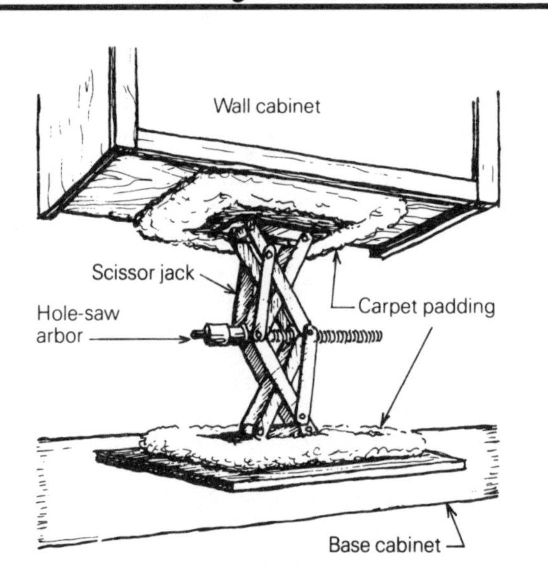

Installing heavy kitchen wall cabinets by yourself can be a dangerous juggling act. With the aid of a scissor jack it's easy to keep them plumb and level long enough to install them. First set the base cabinets. This leaves you with a working surface. Next, lay out level and plumb lines on the wall for the upper cabinets. Place a piece of scrap plywood on the top of the base cabinet and set the jack on the plywood. Then predrill the uppers for the stud layout, set the cabinet on the jack and crank it into position.

To protect the casework from scratches, attach a 12-in. square piece of ¾-in. plywood to the bottom and top of the jack with countersunk flathead screws, and cover the plywood with carpet scraps. I fitted my jack sleeve with an old hole-saw arbor so that I can raise and lower the jack with my electric drill. *—Ron DeLaurentis, North Aurora, Ill.*

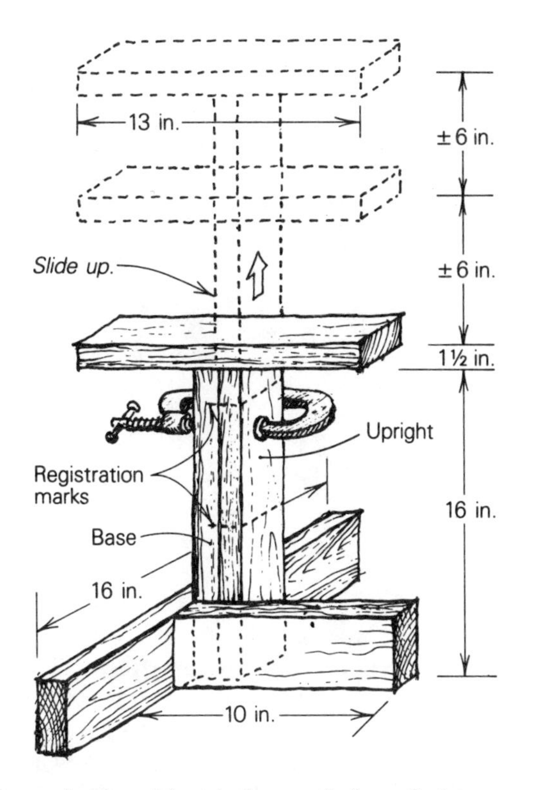

I use these simple cabinet jacks, made from 2x4 scraps, to support kitchen wall cabinets while I adjust their final position. Set on a 36-in. countertop, the jacks will accommodate the three most common sizes of upper cabinets—18 in., 24 in. and 30 in. high.

I use them in pairs, and pre-set each one to the correct rough height registered with marks on the jack base. I use a C-clamp to hold the upright to the base, but a slot in the upright that rides on a carriage bolt through the base would be better. A wing nut could then clamp the upright and base together at any height.

I lift the cabinets onto the jacks, then level and position them using floor-tile scraps as shims. After the cabinets are installed, the jacks are easily removed by releasing the clamp on the upright. —***Jim Barrett, Southwick, Mass.***

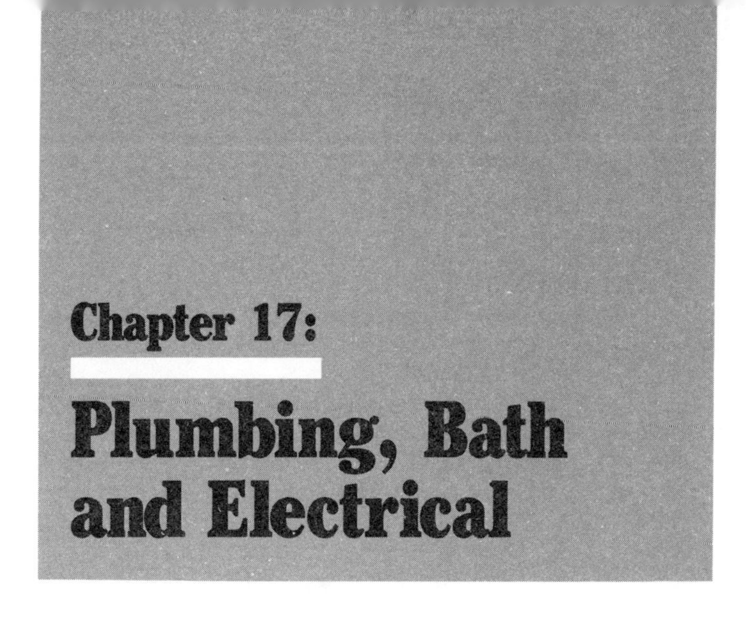

Chapter 17: Plumbing, Bath and Electrical

New soldering in old plumbing

One of the problems I have when remodeling a kitchen or bathroom occurs when I have to sweat-solder new fittings into an existing copper water-supply line. It is almost impossible to get all of the water out of the pipes, and that leads to poor solder joints.

My solution is to drain the pipes in the usual manner—being sure to open faucets at the highest point in the system—and then use my wet-vac to suck the remaining water out of the lines. This procedure requires a vacuum-nozzle reducer to ensure a tight fit and good suction (I use the plastic reducer that connects my orbital sander to its dust bag). Remember to vacuum both the uphill and downhill ends of the line. I find that after about two minutes the pipes are dry and ready to solder.

—Peter D. Ellis, Needham, Mass.

Drainpipe patch

When I've made temporary repairs on exposed, low-pressure drainpipes such as those used for washing machines or swimming pools, I've found automotive radiator hose to be quite useful. Radiator hoses are available in several diameters, and they come in a variety of offset shapes that can fit many situations. Best of all, these hoses are manufactured to withstand high pressures and temperatures, so it is unlikely that any home applications will exceed their capabilities.

To repair a drainpipe, I remove the damaged portion and file smooth any burrs. Then I slip radiator hose over both ends of the pipe, and cinch them with a hose clamp. This allows me to repair broken lines with ease.

—Greg Clarke, Mountain View, Calif.

Cable pulley

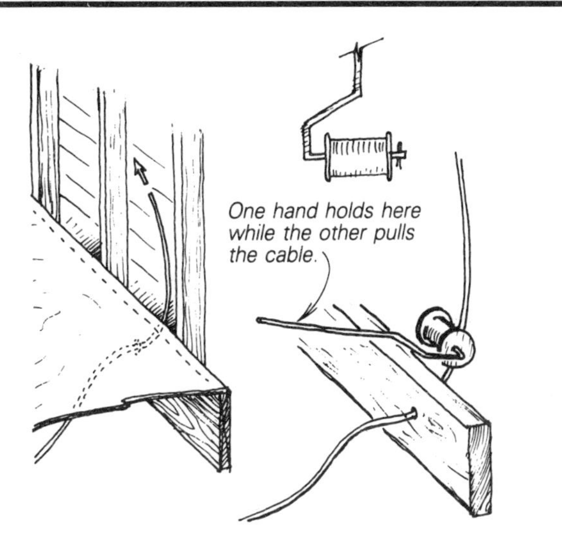

While rewiring my house, I was confronted several times with having to thread the cable behind finished walls and under subfloors. Difficulty arose when the cable had to be drawn through holes that were perpendicular to the only direction I could pull the cable.

To keep the cable from binding against the edges of the hole, I fabricated the contraption shown above. It's a 16-in. piece of ⅜-in. steel rod. At one end is a plastic spool that once held monofilament fishing line. A cotter pin holds the spool in place. Used as a pulley, the spool makes it easy to pull cable through a hole that's at right angles to the direction of pull. ***—Malcolm McKeag, W. Kingston, R. I.***

Wiring in old walls

Here are some tricks to help you use a fish tape when you're trying to run new wiring through old walls. A fish tape is a thin, spring-steel wire with a hook on one end. It's mounted on a spool, and most often used to pull wiring through sections of conduit. It can also snake through hidden wall and ceiling cavities.

It can be hard to hit a target with fish tape. So I catch it with a string. When you're working on a vertical run, such as an attic to a wall-mounted outlet, lower the string from above. It should be tied into a series of loops, one pushed through the other, and the string weighted with a piece of solder. The fish tape is then introduced into the lower access hole, and manipulated until its hook catches one of the loops. The same weight can work horizontally—push the weighted string along the flat run with a stick until it falls into the vertical cavity.

Another way to get a string from one place to another in an empty conduit is to tie it around a small wad of cellophane and use compressed air to blow it down the run. Then tie it to a heavier string and you're ready to haul on the tape. For a really convoluted run, where a normal fish tape won't follow the necessary path, a speedometer cable can do the job. When you are doing this kind of work, make sure that you know the location of any live boxes and unprotected wiring, like knob and tube work. And to make things easier in the event of future wiring changes, leave a piece of nylon string pulled through the runs.

—Norman Rabek, Burnsville, N. C.

Outlet planning

A friend recently came to me ready to wire the home he was building for stereo, cable television and telephone. He wanted outlets in each room, but he didn't know which system would end up where in the long run. I suggested that he choose a couple of logical places in each room, and use a standard outlet box with a blank cover. With all three wiring systems brought into the box and capped, he could choose the system he wished to use and replace the blank cover with the proper outlet cover. This way he wouldn't need to have a lot of exposed wires, or galleries of outlets that weren't being used. ***—L. Fredrick, Aspen, Colo.***

String snare

The time-honored way to connect stereo speakers or TV antennae is to drill a small hole in the floor, and then run the wires across the basement ceiling. That's fine if you've got a basement, but my house has a 2-ft. high crawl space that is a tapestry of cobwebs and spiders. Rather than crawling 12 ft. to reach the hole in the floor, I put my chimney-cleaning brush to work.

I pushed a length of string through the hole so that it dangled below the floor. Then I attached the brush to a 15-ft. long fiberglass pole and snagged the string in its bristles. After a couple of twists, I had the string firmly entangled. After I pulled it in, I used the string to pull a speaker wire to its final destination. ***—Bruce Knott, Grand Haven, Mich.***

On-site bender

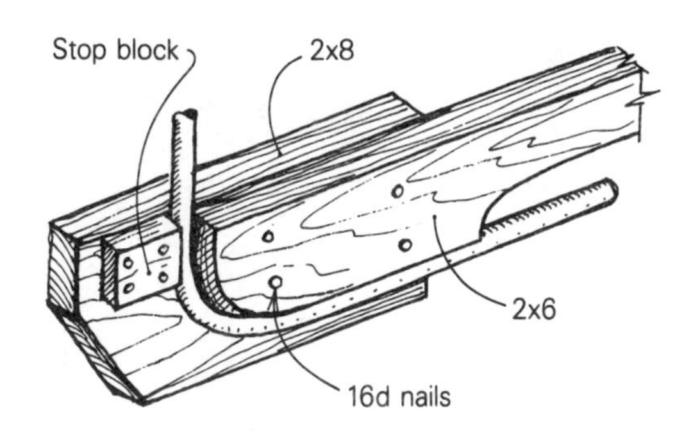

Nothing is worse than showing up at a job to discover that an important tool didn't make it into the truck. One day, I left my conduit bender at home, so I used scrap wood to make a bender that gave me beautiful, kink-free bends in my ½-in. conduit.

I started with a piece of 2x8 about 15 in. long for the body, and I attached a short piece of 1x4 to it with screws to act as a stop. Next I cut a 90° arc into the end of a 5-ft. 2x6, and nailed it to the body as shown above. Since most of the 2x6 needed to act as a handle, I tapered its width to about 3 in. to make the tool less cumbersome, and I was back in business. ***—Richard Tufts, Santa Rosa, Calif.***

Securing electrical tape

Whenever you are using electrical (PVC) tape to wrap wire connections, rope ends or repaired power cords, just swab the end of the tape with PVC pipe cement. This will keep it from coming unstuck and unwinding.

—Bill Hart, Templeton, Calif.

Tub drop-ins

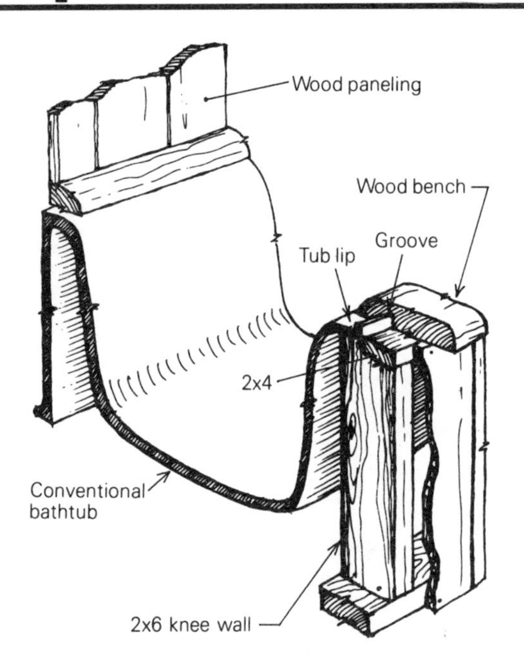

Using materials and fixtures in a way that wasn't originally intended is a mainstay of remodeling work. This can even be true of bathtubs. My project last summer was a tub installation that called for wood paneling around the walls and included a low bench at the side. What I needed was a drop-in tub. However, these are very expensive, hard to color-match to existing fixtures, and take a long time to get after ordering.

My solution was to purchase a standard tub. By installing the finished skirt against the back wall of the tub area, as shown in the drawing, instead of in front as usual, I was left with a completely open outer side for framing the knee wall and applying the seat, which I grooved to fit over the lip of the tub. *—James B. French, Portsmouth, R. I.*

Fixing a leaky toilet

I had a toilet leaking around the rubber stopper at the bottom of the tank, which not only wasted water but caused pure annoyance at the sound of the tank endlessly filling at a trickle. Recalling that the life of rubber skin-diving equipment can be lengthened with applications of silicone spray, I decided to try some silicone on the rubber stopper. I drained the tank, and without removing the stopper I sprayed it with the silicone and worked the preservative into the rubber with my fingers. The spray evidently added some suppleness to the stopper, because when I refilled the tank, the library was once again quiet. Silicone spray is available at auto-supply stores as well as dive shops.

—Stephen Denton, Somerville, Tenn.

Soldering water pipes

An old plumber's trick for resoldering copper water lines when making repairs is to take a piece of white bread, make two dough balls and then stuff them up the pipe in both directions. The dough balls will block the water while the soldering is being done, and then dissolve easily in the water system. ***—Chris Valenzuela, Millboro, Va.***

Plugged-up plumbing

If you're working on an old galvanized water supply system (replacing a section with copper pipe to improve water pressure, for example) and nothing comes out of the faucet when you turn it on, the faucet screen is probably packed with rust particles and flux from the soldering operation.

—Chas. Mills, Leadville, Colo.

Wiring guide

If you come back to a wiring job days or weeks after installing the cables, you sometimes have trouble remembering which wire goes where. I have an installation tip that prevents this problem. I always cut the switched lines on a diagonal, and I square-cut the power leads. ***—Mike Lyon, Tacoma, Wash.***

Buzzless bulbs

A contractor friend of mine recently replaced some of my light switches with dimmers. I like the adjustable light levels, but I hated the buzzing noise made by the bulbs when they are dimmed. After some research, I discovered the solution—three-way bulbs. Their filament construction is such that they don't buzz when dimmed.

—Mary Jacek, Alameda, Calif.

Rotary fishing rod

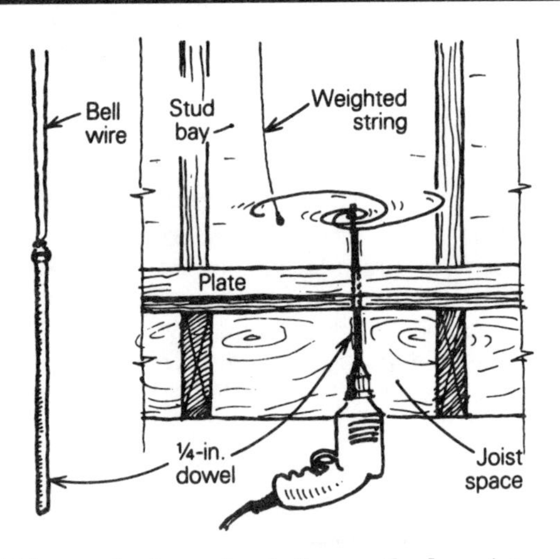

When fishing a wire for a doorbell recently, I ran into a seemingly impossible situation. I'd drilled the hole for the bell-push, and an angled hole through the wall plate into the proper stud bay from the cellar. But try as I might, I couldn't get a wire from one to the other. I tried fish tape, a weighted string, bell wire, bead chain, profanity, hooks, probes and a dozen other ploys. I couldn't find any obstruction, but I couldn't find the wire either.

As my last attempt before starting to rip clapboards off the side of the house, I made the "fishing rod" shown in the drawing. Starting with a piece of ¼-in. dowel about a foot long, I drilled a ⅟₁₆-in. hole across the diameter about ⅛ in. from one end. Through this hole I inserted a 15-in. piece of bell wire that I secured with a square knot, leaving the two ends equal.

I then chucked the other end of the dowel in an electric drill. Folding the ends of the wire so they stuck out ahead of the dowel like antennae, I shoved the contraption into the hole in the plate as far as it would go and turned on the drill. At 1,200 rpm, the ends of the wire whipped out centrifugally, lashing around inside the wall and almost immediately entangling the weighted string left dangling from the bell-push hole. When I pulled the drill back through the hole in the plate, I found the string securely wrapped around the dowel. ***—Robert L. Edsall, New Haven, Conn.***

Circuit savvy

While remodeling my house, I have found it frustrating trying to locate the breaker that deactivates the circuit I am working on. The circuit box is no longer labeled correctly, and won't be until I finish. So once I match a breaker with its circuit, I label the circuit-breaker number on the inside of the outlet cover plate with a grease pencil. Now when I have to work on an outlet, I just unscrew the plate and check it for a number—then I go downstairs and flip the right breaker.

—L. D. Frederick, Aspen, Colo.

Fitting drywall

As a remodeler, I hang and tape a lot of drywall. To get a close fit around electrical outlets, I use the installed outlet itself to mark the gypboard. Spread liquid detergent or hand lotion around the rim of the outlet box, and lift the drywall panel into place on the wall, pressing it tightly against the outlet box. The back paper will absorb the liquid immediately upon contact, marking the outlet clearly. Then just cut around the outline for a tight fit.

—Mark Messier, Eugene, Ore.

Tub support

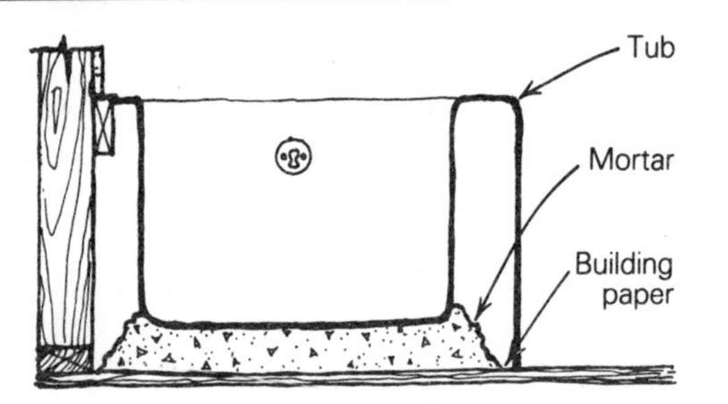

If you are about to install a lightweight steel or fiberglass tub, here's a way to make it feel more substantial. Set it on a bed—or rather a blob—of mortar.

First staple building felt to the subfloor in the area to be occupied by the tub. Spread a thick, stiff blanket of mortar onto the felt so that it will cradle the bathtub, as shown in the drawing above. What was once a flimsy tub bottom with no thermal mass is now solid as a rock and able to retain warmth. It's a very nice touch if you can't use cast iron.

—Kurt Lavenson, Berkeley, Calif.

Wonderboard rasp

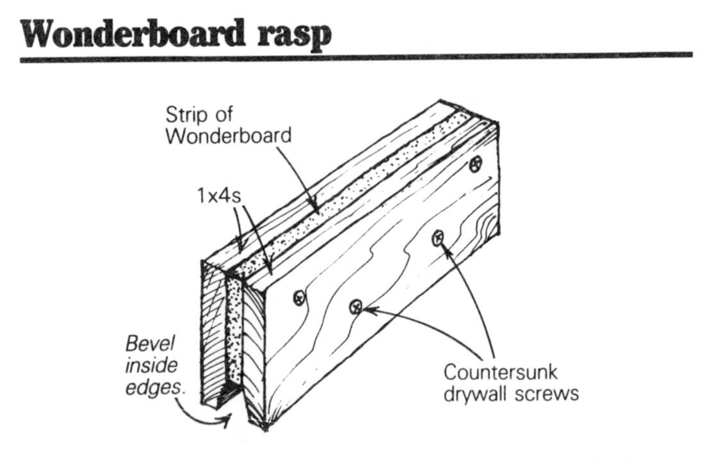

I built a complicated bathroom last year that included a lot of tile. As a setting bed, I used a tile backing panel called Wonderboard. This ½-in. thick cement and aggregate panel is faced with a fiberglass mesh. To cut it, one scores the mesh (on both sides is best). Then the panel will break in a manner similar to drywall, and like drywall the resulting edge will be ragged.

Because of the bathroom's complexity, the Wonderboard edges had to be as close to perfect as possible. I've used a Surform plane to smooth a ragged drywall edge, but Wonderboard is so tough that a metal-edge tool would be dulled instantly by it. The tool in the drawing on the facing page is my solution to this problem. It uses a strip of Wonderboard as a cutting edge. I sandwiched the strip between a pair of 1x4s, and held the pieces together with drywall screws. To ensure friction-free travel along the edge of the Wonderboard, I beveled the inside edges of the 1x4s.

—Mark S. Goldman, Boston, Mass.

Shower-stall backing

Here's an inexpensive way to make the thin, flimsy walls of a fiberglass shower more rigid. Apply expanding urethane foam at several spots up and down the wall studs behind the shower walls, as shown in the drawing. The foam will do its expanding sideways. When it cures, the shower walls will have a lot less give. ***—Byron Papa, Schriever, La.***

The bathroom angle

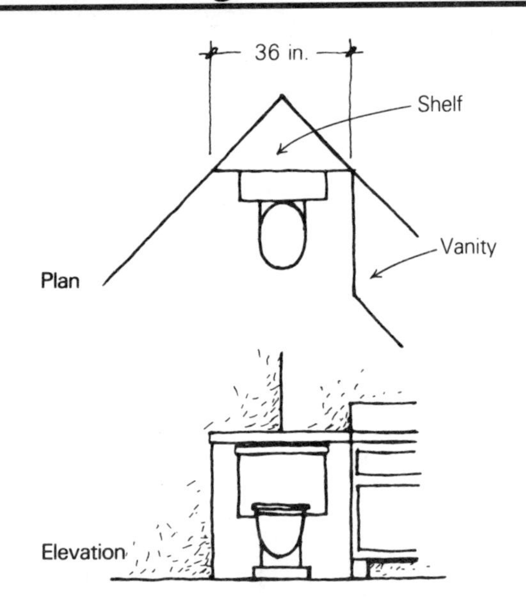

The drawing above illustrates an idea that I use frequently in the houses that I design. It revolves around mounting the toilet at a 45° angle in the bathroom corner. Positioned this way, the toilet looks better, it's easier to clean around and there is room behind it for a convenient shelf.

These were some of the bonus features of the design. My original intention was just to make it more convenient for folks in wheelchairs. Many have to face the toilet to get on it...me for instance. —***Jay Wallace, Ashland, Ore.***

Grout gun

We bought a house with a nicely remodeled bathroom. All it lacked to make it look really finished was a tile baseboard.

With the sink, toilet and other fixtures still in place, I was able to spread the mastic to hold the tiles to the bathroom wall with a fair amount of contortion but no real trouble. The only problem I foresaw was with the unwieldy rubber grout trowel—there simply wasn't enough room to maneuver it in some of the tight places.

Casting about for an alternative grouting method led me to the kitchen drawer and the seldom-used cake-decorating device. It looks like a caulking gun, with several interchangeable metal tips and a piston plunger at one end to control the flow of icing, or whatever. The cap unscrews for loading. I filled the thing with grout and went to work.

The device worked remarkably well for this application. I was able to reach remote crevices with ease and fill them with a high degree of accuracy.

My new tool was easy to clean when I was done with it, and I returned it to its kitchen duties none the worse for wear.
—Chuck Gomez, Ponca City, Okla.

Beefed-up towel bars

The porcelain towel-bar fixtures that are available around here have attractive ceramic brackets, but the towel bars are hollow, flimsy 1-in. square plastic tubes. To make the bars as sturdy as their brackets, I fill them with plaster of Paris.

First I plug one end of a bar with a lump of plumber's putty. Then, with the bar braced and standing, I fill the tube to within about 1 in. of the top with a thick slurry of plaster. When the plaster hardens, the bar has a solid, rigid feel that is in keeping with ceramic fixtures. ***—Dennis LaMonica, Panama, N. Y.***

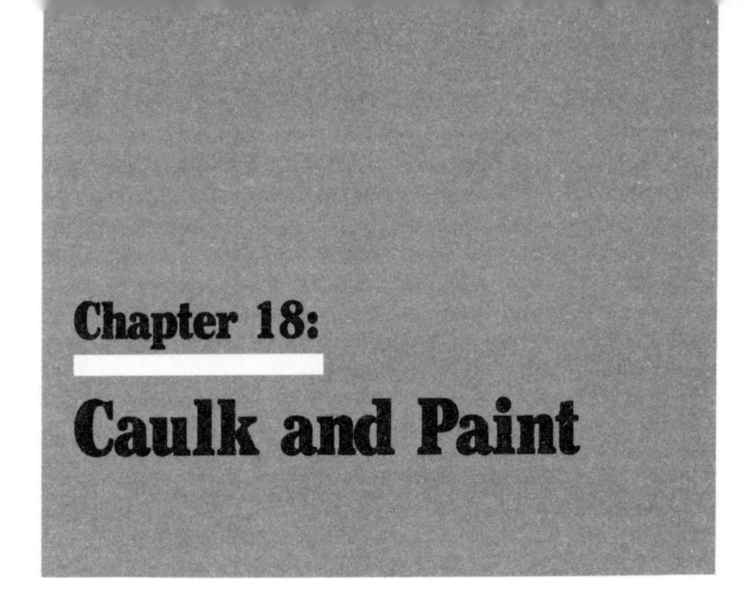

Chapter 18: Caulk and Paint

Caulk trimmer

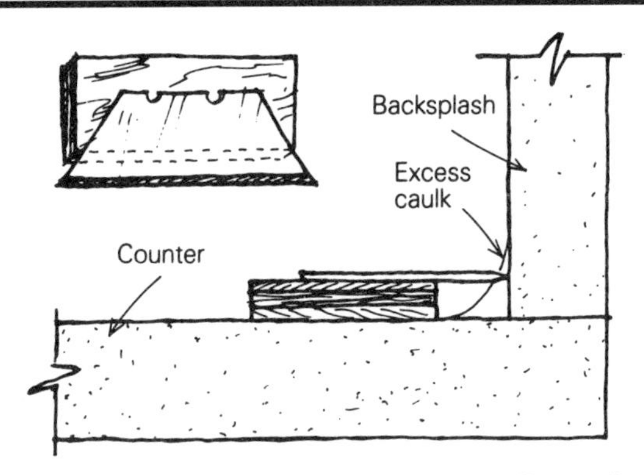

When I recently installed a new kitchen, my client asked me to caulk the seam between the counter and the backsplash rather than cover it with a piece of metal trim. First I roughed up the corner with sandpaper, and then I applied a small bead of silicone caulk. Uncertain that the caulk was tightly compacted into the corner, I gave into temptation and smoothed it out with my finger.

I allowed it to dry overnight, and the next morning I trimmed it back to a uniform dimension using the tool shown in the drawing. The tool consists of a sharp razor blade glued to a small piece of 1/4-in. plywood. I ran the plywood along the counter and backsplash, then gently lifted the excess off with another razor.

—*Pulin Sumner, Jamaica, N. Y.*

Applying liquid preservatives

Many builders treat exterior millwork with liquid preservatives (usually penta or chromated copper arsenate) to protect the wood against fungal decay. Unfortunately, chemicals that kill fungi can also be deadly to humans, animals and plants. Spraying is a hazard because of drifting mist, and dip-treating in vats is practical only for small pieces or commercial operations. Brushes or rollers can't hold much of these thin, volatile liquids, so more preservative dribbles down our sleeves than onto the woodwork. The challenge, then, is to treat the wood without spilling any on our skin, clothing or the surrounding area.

To apply liquid preservatives, I use a plastic squeeze bottle that in a previous life contained liquid dish soap. I squeeze the empty bottle to expel air, dip the tip into a can of preservative, and release my grip so the bottle sucks in the liquid. With the bottle about half full in one hand, I feed a brush (or roller) in my other hand. I squeeze preservative directly onto the bristles to keep them wet while I move the brush along the wood. If liquid starts to trickle, I stop the flow, catch the drips with the brush, and work the excess into the surrounding wood. This way I can even control the flow of preservative while treating corners, edges and vertical surfaces. Of course, I still wear rubber gloves and goggles, but with this method more preservative ends up on the wood and less on me and other living things.

—*J. Azevedo, Corvallis, Ore.*

Spud fingers

If you are caulking a joint and running out of clean fingers to smooth the caulk with, try using a potato. Before a job, I carve spuds to the desired profile, and store them in a plastic bag. I've found that the acidic potato juice keeps silicone and other caulking materials from sticking to the spud fingers.

—*Glenn Zahn, Fairbanks, Alaska*

Drip-catching cuffs

In a kinder world, we would always be able to apply wood finishes to horizontal surfaces at bench height. But in real life we invariably find ourselves reaching over our heads with a brushful of stain or oil. So you know as well as I the misery of cold, sticky fluid dribbling down over your wrist. Next time try this trick.

Get a rubber glove, of the type sold for washing dishes, for your painting hand. Turn up a 2-in. to 3-in. cuff on the glove, and stuff it with toilet paper—you want a puffy doughnut of tissue filling the cuff and circling your wrist. The tissue holds the cuff out to catch the dribbles and absorbs the fluid so it won't leak out when you lower your arm.

Finally, wrap the end of the brush handle with tissue and secure it with a rubber band. You are now ready to apply the finish. When the tissue rings become saturated, squeeze them out over the finish container to put the liquid back where it can be used again.

—Jerry Azevedo, Friday Harbor, Wash.

Varnish stilts

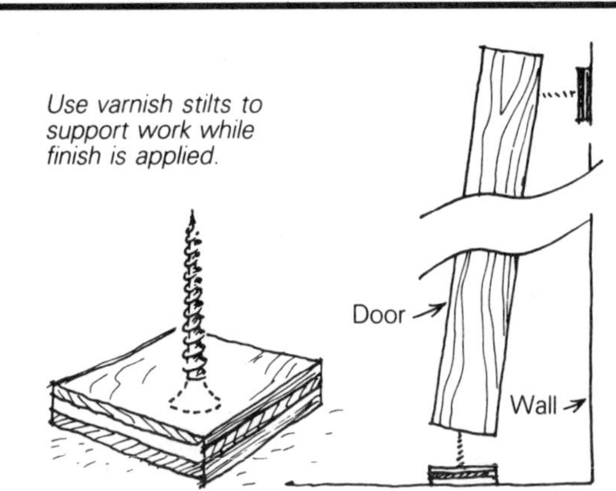

Use varnish stilts to support work while finish is applied.

A drywall screw through a small square of plywood makes a handy stilt to prop up work while finish is applied. When I paint a door, for instance, I place a pair of stilts on the floor to support the door, and another one against the wall near the top of the door, as shown above.

—Michael R. Sweem, Downey, Calif.

Heavy-duty paint scraper

I had been puzzling for some time over a safe and efficient way to remove the peeling paint from our propane tank, when our dog solved the problem for me. The dog had carried off the horse's curry comb, and left it a few feet from the tank. As I picked it up, it crossed my mind that it might make a good scraper. It did. In five minutes, I had removed all the loose paint from the tank. Since then I've used the curry comb as a scraper on other repainting projects. I've found it to be faster than ordinary scrapers, and its handle makes it easier to hold without tiring. ***—Betty Lartius, Ames, Iowa***

Caulk-tube nozzles

I've been a siding contractor for ten years, and I'm continually confronted with two conditions that are difficult to caulk properly. One is when I have to get a good bead of caulk along the trim over a door or window that is close to a soffit—there isn't enough room for the caulking gun. In this situation, I tape a 6-in. length of ⅜-in. ID vinyl tubing to the caulk-tube tip. I use duct tape for this because it's strong enough to resist the pressure of the caulk as it's forced out of the tube. The flexible vinyl tubing can be bent into the right position with one hand, while I work the gun with the other.

The other problem is caulking around meter boxes that are too close together to allow the tube between them. In this case, I attach a plastic drinking straw to the end of the tube with duct tape. The straws are rigid, and long enough to place the caulk where I want it. ***—Bill Bugajski, Chicago, Ill.***

Poly paint tray

On a job where painting is done by roller before the finish floor is installed, save the expense of disposable roller-tray inserts or the hassle of cleaning the tray itself. Instead, cut a 3-ft. square piece of 6-mil polyethylene and spread it out in the middle of the room. You can pour almost half a gallon of good-quality latex paint on the plastic, and fill your roller from this central well. When you are finished for the day, cut the plastic in half and wrap it around the roller to keep the roller fresh for the next day—no need to rinse out the paint-laden roller. Use a fresh square of poly every day, or you will have flakes of dried paint marring the work.

—Rick Lazarus, Spencer, N. Y.

Caulk cap

There are a lot of new caulks and adhesives available in tubes these days, but many tasks need only a little squirt of the high-priced goo. Left-over tubes might just as well be tossed into the trash on the spot because the material in the tip becomes hard and unusable, rendering the rest of the contents inaccessible. An effective cap would solve the problem—and I had some in my tool box all along.

The solution is wire nuts, the vinyl-coated red ones that are normally used on 12-ga. wire. I've found them effective for up to six months, and they can be constantly reused and easily replaced. *—Joe Shepherd, Milwaukie, Ore.*

Paint caddy

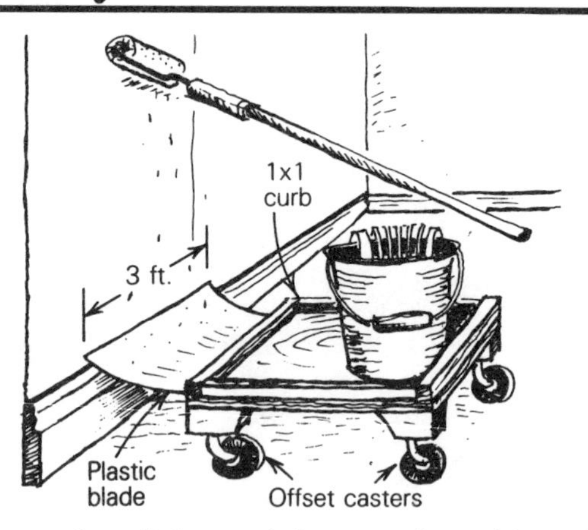

I paint houses for a living, and there are three things about painting with a roller that really annoy me. One is masking the baseboard and spreading out a tarp to catch drips and speckles. Another is moving the paint bucket from station to station, and the third is moving the tarp again, which inevitably results in drops of wet paint smearing the carpet or floor.

I now use a plywood platform on wheels both to carry the paint bucket and to protect the baseboard and floor from paint splatters. A plastic blade on one end of the platform butts up against the wall to catch drips. To make the blade, I used a section cut out of an old plastic garbage can, and I attached it to the caddy so the curve points upward. This directs paint away from the wall, and accommodates the higher baseboards. *—Stan Lucas, Redmond, Wash.*

Scraper cleanup

Paint stripper combined with old paint or varnish makes a sticky goo that can be tough to remove from a scraper or a putty knife. To make an easy job of it, I cut a straight slit about 2 in. long in a large tin can. Then I slide the blade of the knife into the slit close to where it joins the handle. When I pull the blade out, the old finish falls into the can, ready for disposal. *—Roy Viken, Boise, Idaho*

More life for sanding belts

Resin or paint can quickly clog up a perfectly good sanding belt. When this happens, I flip the sander over, and without removing the belt, I scrub off the gummy stuff with a file card, a wire brush made for cleaning files. Its stiff, short bristles are perfect for cleaning belts. When a belt that's still clean shows signs of dullness, it's lost some of its abrasive grains, and those remaining have had their leading edges broken down and rounded off. Before discarding the belt, turn it around. The grains are dulled in only one direction, and another 10% more use can be gained before the grains are dulled again or the seam separates. *—Philip Zimmerman, Berkeley, Calif.*

Paintbrush carrier

When I paint trimwork, I hate to waste time climbing up and down the ladder to get the right brush. I could leave my various brushes in my paint bucket, but they would soon become a dripping mess.

My solution is to cut the top off a one-gallon plastic antifreeze jug with flat sides. I then cut two slits in one of the sides and thread a nylon belt through them, as shown in the drawing. I slide the loose belt ends through the loops of my painter's pants, creating a paintbrush carrier that rides easily on my hip. I now have several brushes close at hand. The carrier cleans up easily with water or paint thinner.

—Mike Ellis, Seattle, Wash.

Cleaning paintbrushes

I've been a builder for many years, involved in all phases of construction, and at some point in nearly every job I've had to get out the paintbrushes. I admit I'm no artist, but I have learned a few techniques along the way.

When I purchase a new brush, I get a pure-bristle one with an unpainted handle—there's no store-bought finish to peel off the raw wood, and the oil from my hands preserves the handle. I usually put a small nail above the ferrule and bend it down toward the bristles at a 90° angle. This allows the brush to hang in a can of paint or thinner without touching bottom, so the bristles won't warp.

When it comes time to paint, I take a 16d nail and punch a ring of holes in the deep part of the groove at the top of the opened paint can. These holes allow the paint that accumulates when you wipe the excess from your brush to drip back into the can, rather than overflow and run down the outside. When the lid is replaced, the holes are sealed inside.

Whenever I use a brush with oil-base paint, I clean it with paint thinner and liquid laundry soap. First, I pour about ½ cup of thinner into a pan, and I work the brush back and forth in it to remove most of the paint. Next, I pour this first batch of thinner into a storage container (a 3-lb. coffee can is fine). Then I rinse the brush in ¼ cup of thinner. I work the bristles with my fingers to get out as much paint as possible, and return the thinner to storage. In the container, the paint solids will settle to the bottom, allowing the clear liquid on top to be poured off and used again.

At this stage in the brush cleaning, I pour about a tablespoon of laundry soap into a pail, wipe it up with the brush and work it into the bristles with my fingers. Next, I fill the pail with water and rinse out both brush and bucket. I repeat this soap sequence one more time, and then I shake out the brush thoroughly and return it to its jacket to make sure that its bristles remain straight. Using this cleaning method, I've been able to use the same brush to apply both paint and lacquer, with no paint residue spoiling the clear finish. ***—Chris Thyrring, Halcyon, Calif.***

Stack painting

Stack boards to paint edges. Stagger them to brush out excess finish.

When you've got to brush paint or stain on the edges of many boards, try stacking them flat so that you can apply finish to all of them at once. This will save you a tremendous amount of time compared to treating each edge separately. When you've got them all coated, slightly stagger the pile and brush out any excess finish that may be on the faces and backs of the boards.

—Ed Kobus, Silver Spring, Md.

The kerosene alternative

Kerosene, sold as heater fuel for $1 to $2 per gallon, makes a good substitute for paint thinner, which costs considerably more. Used as a solvent for brushes, kerosene leaves a slight residue on the bristles which helps to keep them supple during storage. This residue can be removed with mineral spirits if necessary. Kerosene will significantly extend the drying time of oil paints, so keep this in mind if you are using it as a thinning agent.

—Albert Treadwell, Ridgefield, Conn.

Caulking-gun care

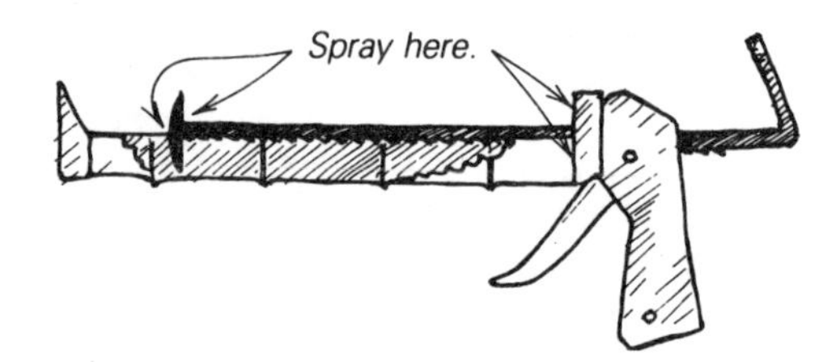

My caulking guns used to accumulate gobs of hardened caulk and mastic on the plunger, making cartridge removal very difficult. The tough rubbery mass (silicone is the worst) was a real nuisance, and prompted the early retirement of several guns. Then I discovered the trick of spraying some silicone lubricant (WD-40 works too) on the plunger and into the back of the barrel. The lubrication lets the encrustations peel off in one piece, and makes changing tubes easy. You'll have to renew this lubricant every so often. One warning, though: If your gun has a friction-type drive mechanism, don't spray it—the lubrication will render it useless. ***—William H. Brennen, Denver, Colo.***

Caulking in tight spots

A piece of plastic sheathing from a 12- or 14-ga. wiring cable (such as Romex) makes an effective caulking-tube extension when the tube's nozzle doesn't have the flex to reach the spot in need.

—Malcolm McDaniel, Berkeley, Calif.

Painted-trim prep

Sometimes when a house interior needs repainting, the woodwork is too rough to take a new coat of paint, but not bad enough to warrant stripping. Under these circumstances, I prepare the woodwork by using a sponge, a bucket of water, a scrub brush and some wet/dry sandpaper. I keep the sandpaper and work surface wet throughout the process, and I use my hands or the sponge as a sanding pad. I find that the paper will last up to an hour if I periodically use the brush to scrub the paper in the bucket.

Using this wet process keeps the dust suspended in the water rather than in the air. The resulting surface is quite smooth to the touch, but it still has the necessary roughness for good paint adhesion. I've used this technique with old oil-base and water-base paints with equally good results. ***—John Glenn, Brookline, Mass.***

Centrifugal roller cleaner

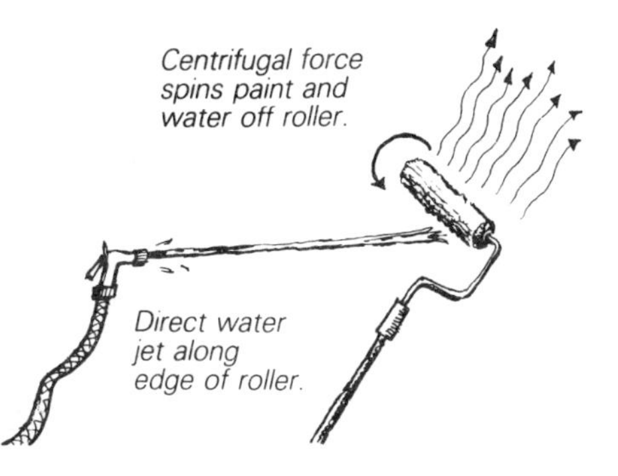

Cleaning water-soluble paint from a paint roller used to be a tedious chore until I came up with this idea. Now after the painting is finished, I simply attach an extension handle to the roller, step outside and use a garden hose to do the work. By directing a water stream along the edge of the roller, the roller revs up to a good speed and spins the paint and water off the roller, as shown in the drawing above. Be sure that you perform this operation well away from anything that might be damaged by the paint and watery overspray.

—Mel Wolpert, Weatogue, Conn.

Index

A

B

C

D

E

F

G

H

I

R

S